100% Renewable

100% Renewable

Energy Autonomy in Action

Edited by
Peter Droege

publishing for a sustainable future

London • Sterling, VA

First published by Earthscan in the UK and USA in 2009

ISBN: 978-1-84407-718-2

Typeset by FiSH Books
Cover design by Rogue 4 Design (concept by Anis Radzi)

For a full list of publications please contact:

Earthscan
Dunstan House
14a St Cross St
London, EC1N 8XA, UK
Tel: +44 (0)20 7841 1930
Fax: +44 (0)20 7242 1474
Email: earthinfo@earthscan.co.uk
Web: **www.earthscan.co.uk**

22883 Quicksilver Drive, Sterling, VA 20166-2012, USA

Earthscan publishes in association with the International
Institute for Environment and Development

A catalogue record for this book is available from the British Library

Library of Congress Cataloging-in-Publication Data
100 per cent renewable : energy autonomy in action / edited by Peter Droege.
  p. cm.
Includes bibliographical references and index.
ISBN 978-1-84407-718-2 (hardback)
1. Renewable energy sources. I. Droege, Peter. II. Title: One hundred per cent renewable.
TJ808.A15 2009
333.79'4—dc22
        2009018125

At Earthscan we strive to minimize our environmental impacts and carbon footprint through reducing waste, recycling and offsetting our CO_2 emissions, including those created through publication of this book. For more details of our environmental policy, see www.earthscan.co.uk.

This book was printed in the UK by
CPI Antony Rowe.
The paper used is FSC certified.

Mixed Sources
Product group from well-managed
forests and other controlled sources
www.fsc.org Cert no. SGS-COC-2953
© 1996 Forest Stewardship Council
FSC

Today I challenge our nation to commit to producing 100 per cent of our electricity from renewable energy and truly clean carbon-free sources within ten years. This goal is achievable, affordable and transformative.

Al Gore, Constitution Hall, Washington DC,
17 July 2008

This is absolutely nothing to do with emissions trading, this is about getting on and doing it. You cannot tackle climate change by trading, you have to actually do things.

Allan Jones, London Climate Agency, about his work for
Woking Borough Council, on Australia's ABC Radio National,
26 July 2008

We will harness the sun and the winds and the soil to fuel our cars and run our factories.

President Barack Hussein Obama, inauguration speech,
20 January 2009

Our survival will very much depend on how well we were able to draw down CO_2 to 280 parts per million.

Hans-Josef Schellnhuber, Potsdam Institute for Climate Impacts
Research in Germany, quoted by Gaia Vince in the
New Scientist,
25 February 2009

Contents

List of Figures, Tables and Boxes

Figures

Tables

Boxes

List of Contributors

Brian W. Bush
Brian W. Bush is a Principal Strategic Analyst at the National Renewable Energy Laboratory in Golden, Colorado. His areas of expertise include energy and infrastructure modelling, simulation and analysis, high-performance computing, software architecture, design, implementation and testing, discrete-event and continuous simulation, statistical analysis, and geographic information systems. He holds a PhD in physics from Yale University and a BS in physics from the California Institute of Technology. His prior work experience was as a technical Staff Member at the Los Alamos National Laboratory.

Ted Caplow
Formerly a clean energy consultant for Capital-E, LLC, Dr Caplow worked on energy efficiency and carbon offset credits for a range of clients including the California Energy Commission and the US Department of Energy. His expertise also includes water contaminant dynamics, and he has published articles on this topic in the *Journal of Environmental Engineering, Environmental Science & Technology* and *Estuaries.* Founder and Technical Director of BrightFarm Systems, Dr Caplow developed the master plan for the Science Barge, a prototype sustainable urban CEA facility. His subsequent design work on rooftop and facade-mounted agriculture has been published in *Science, Sustain, Urban Land* and *Good.* Dr Caplow holds a BA from Harvard University, an MSc in Mechanical Engineering from Princeton University, and a PhD in Environmental Engineering from Columbia University, where he has taught a class on green buildings.

Nancy Carlisle
Nancy is the Manager of the Integrated Applications Group in Deployment and Industrial Programs at the National Renewable Energy Laboratory (NREL) in Golden, Colorado and a licensed architect in the state of Colorado, USA. At NREL, she led the effort to develop a 25-year master site plan for future build-out of the NREL sustainable campus. She led the effort to get NREL's Science and Technology facility certified as LEED Platinum, the first federal building receiving this designation. She is currently on the project team overseeing the design of a 740-person 'zero-energy' office building for the NREL campus. She has served as the NREL lead for the DOE Federal Energy Management Program for a number of years, working with federal agencies to meet their federal goals for renewable energy use at their sites. She has managed a strategic initiative for NREL on the topic of renewable energy

communities. She has worked at NREL for over 25 years, in research, analysis, design and outreach activities that promote the design of sustainable low energy buildings. She is a LEED accredited professional and recognized as a Fellow in the American Solar Energy Society. She holds a Masters degree in Architecture, a Masters degree in Urban Planning and a Bachelors degree in Economics with a concentration in Environmental Studies.

Peter Droege

Peter Droege is Chair of the World Council for Renewable Energy for Asia Pacific, Professor of Sustainable Development at the University Hochschule Liechtenstein and Conjoint Professor at Australia's University of Newcastle. He serves on a number of international panels, including the Expert Commission, Cities and Climate Change of the World Future Council and the Steering Committee of the Urban Climate Change Research Network. As principal at Epolis.com.au, the Sydney-based environmental design firm, Peter Droege has been an executive adviser to numerous government, state, local government and private corporate institutions on renewable and sustainable urban development strategies. His academic background includes the Technical University of Munich and Massachusetts Institute of Technology, and professorial positions at Tokyo University and the University of Sydney. Droege has authored *The Renewable City: A Comprehensive Guide to an Urban Revolution* (Wiley) and edited *Intelligent Environments and Urban Energy Transition: From Fossil Fuels to Renewable Power* (both Elsevier).

Hans-Josef Fell

Mr Fell has been a member of the German Bundestag since 1998 and is spokesman for the Alliance 90/The Greens parliamentary group on energy and technology, and also the spokesman for Technology Assessment. Hans-Josef Fell was the author of the proposal of the German renewable energy act (EEG). He also initiated new research programmes for renewable energies. Fell serves as Deputy Spokesman of the Bavarian Solar Initiatives; Vice-President of Eurosolar; member of the World Council for Renewable Energy (WCRE); Initiator of the influential Energy Watch Group. Awards include: Eurosolar German Solar Prize 1994; prize of the German section of the International Solar Energy Society (DGS) in 2000; Energy Globe Award in 2000; Nuclear-Free Future Award in 2001; German Solar Industry Prize in 2002; first German Geothermal Award in 2002; German Biogas Prize in 2003.

Dieter D. Genske

Dr Dieter D. Genske studied Civil Engineering and Geology in Germany and the US and started working on environmental problems, as Humboldt-research fellow in Kyoto, Japan. At Deutsche Montan Technologie (DMT) he became involved in large urban remediation projects, including the International Building Exhibition IBA Emscherpark. In 1994, he was appointed Professor of Engineering Geology at Delft University of Technology and three years later

Professor of Environmental Engineering at the Swiss Federal Institute of Technology in Lausanne, where he founded the Laboratory of Ecotechnics, the first of its kind in Switzerland. He chaired this research laboratory until the end of 2000. He has conducted a number of development projects in low-income countries. Presently, he teaches at Nordhausen University (Germany) and is chairing the Anthroposphere Dynamics Group at the Institute of Environmental Decision, ETH-Zurich.

Cord Hoppenbrock

Cord Hoppenbrock, Dipl.-Geogr., majored in Economic Geography, at the University of Osnabrück, with a minor in Business Studies and Ecology. He gathered work experience at the Institute of Ecological Economy Research and the Institute for Energy and Environmental Research in Heidelberg. Since October 2007 he has been a research assistant and PhD student at the University of Kassel and at deENet, deENet.org.

David Jacobs

David Jacobs is researcher and PhD candidate at the Environmental Policy Research Centre in Berlin (FFU). His research focuses on support mechanisms for renewable electricity. Previously, he gained work experience in the German Bundestag and large European utilities. He also worked on projects for several research institutes, the German Wind Energy Association and the World Future Council. David Jacobs is a frequent speaker at international conferences. He has an academic background in International Business and Cultural Studies.

Walter James

Wal James is a mechanical engineer, graduated from the University of Santiago, Chile, and is a member of the Institution of Engineers Australia. His research and development experience is manifest in his role as Associate Researcher at Curtin and Murdoch Universities in Perth, Western Australia, investigating renewable energies and their connection with plug-in hybrid electric vehicles. He has been a Board Member of the Cooperative Research Centre for Renewable Energy Australia, Projects Manager Murdoch University Energy Research Institute, and Projects Manager Solar Energy Research Institute of Western Australia. His industry experience is extensive: Wal James was founding director of Advanced Energy Systems, and served as chief executive officer of Energy Efficient Co., Singer Controls and RCA Arica, Chile.

Lioba Kucharczak

Lioba Kucharczak Dipl.-Ing. agr. studied agricultural sciences with a specialization in Environmental Management and Development of Rural Areas. Since January 2008 she has been part of the project team 'Sustainable 100% renewable energy regions in Germany' at deENet.org and a PhD student at the Department of Grassland Science and Renewable Plant Resources at the University of Kassel.

Stefan Lechtenböhmer

Stefan Lechtenböhmer Dr rer. pol. is Co-Director of the Research Group Future Energy and Mobility Structures of the Wuppertal Institute for Climate Environment Energy, Germany. He is responsible for the applied research in national and international sustainable energy and climate scenario analysis. He has conducted numerous studies on energy and emission scenario analysis and pathways to a low-carbon society. A particular focus of his work is the design and evaluation of energy efficiency policies and measures. His further research topics comprise greenhouse gas emission inventories and projections, sustainable urban infrastructures, and greenhouse gas emissions from the natural gas lifecycle. He acquired his PhD in Energy and Environmental Management at the International Institute for Management at the University of Flensburg. He holds a university degree (diploma) in Geography, Economy and Political Sciences from the University of Münster and is a member of the UNFCCC Roster of Experts for Greenhouse Gas-Inventories, Policies & Measures, Greenhouse Gas-Projections.

Harry Lehmann

Dr Harry Lehmann works for the German Federal Environment Agency as Head of Division I: Environmental Planning and Sustainable Strategies. From 1983 to 1991 he served as Head of the Consultancy 'UHL Data' – Systemanalyse und Simulation im Bereich Energie- und Umwelttechnik. Until 2000 Lehmann was with the Wuppertal Institute for Climate, Environment and Energy, and until 2006, he was Vice President of Eurosolar. Between 2000 and 2002 he was appointed Member of the Enquete Commission 'Sustainable energy supply under the boundaries of globalization and liberalization' of the German parliament. A founding member of the World Council for Renewable Energy he served as Solutions and Innovations Unit Director for Greenpeace International. Also in the early 2000s he headed the Institute for Sustainable Solutions and Innovations in Aachen. Since 2004 Dr Lehmann has been General Director at the Federal Environment Agency of Germany and President of the Factor 10 Club.

Miguel Mendonça

Miguel Mendonça is Research Manager for the World Future Council. His background is in horticulture, geography, history, journalism, social science and environmental ethics. He is a researcher, writer and advocate, focusing on renewable energy. He has worked on four continents, campaigning, coalition-building and speaking, and is a member of the steering committee of the Alliance for Renewable Energy, promoting feed-in tariffs in North America. He writes books, articles, papers, comment pieces and reviews on sustainability issues, is author of *Feed-in Tariffs: Accelerating the Deployment of Renewable Energy* and is co-writing a new book on decarbonizing the global economy, entitled *A Renewable World – Policies, Practices, Technologies*.

Peter Moser
Dr Peter Moser was project manager at the Centre for Environmental Research, Leipzig (1998–2003) and at the University of Kassel (2003–2007). Since 2007 he has served as project manager at the deENet (deENet.org), especially on the research project 'Sustainable 100% renewable energy regions in Germany'.

Peter Newman
Peter Newman is the Professor of Sustainability at Curtin University and has recently been appointed to the Board of Infrastructure Australia that is funding infrastructure for the long-term sustainability of Australian cities. In 2001–2003 Peter directed the production of Western Australia's Sustainability Strategy in the Department of the Premier and Cabinet. It was the first state sustainability strategy in the world. In 2004–2005 he was a Sustainability Commissioner in Sydney advising the government on planning issues. In 2006–2007 he was a Fulbright Senior Scholar at the University of Virginia Charlottesville where he completed two new books: *Resilient Cities: Responding to Peak Oil and Climate Change* and *Cities as Sustainable Ecosystems*. In Perth, Peter is best known for his work in saving, reviving and extending the city's rail system. Peter invented the term 'automobile dependence' to describe how we have created cities where we have to drive everywhere. For 30 years, since he attended Stanford University during the first oil crisis, he has been warning cities about preparing for peak oil. Peter's book with Jeff Kenworthy, *Sustainability and Cities: Overcoming Automobile Dependence*, was launched in the White House in 1999. He was a Councillor in the City of Fremantle from 1976–1980, where he still lives.

Stefan Peter
Dipl. Ing. Stefan Peter studied Energy and Environmental Technologies, with a special emphasis on renewable energies, at the Fachhochschule Aachen, Germany. The focus of his later work has been on energy efficiency, renewable energy support measures, the possible contribution of renewable energies to energy supply and the integration of renewables into existing energy supply systems. He has authored studies for government departments and independent renewable energy organizations. As one of the main contributors to the 'Energy Rich Japan' study he is familiar with energy system simulation and design and scenario development.

Robyn Polan
Robyn studies Sustainable Development at the University of New South Wales. Born and raised in Toronto, Ontario, she received a Bachelor of Science at Dalhousie University in Halifax, Nova Scotia. Robyn contributed research to Chapters 6, 11 and 13 in this book.

Lars Porsche
Since 2002, Lars Porsche has worked for the Federal Office for Building and

Regional Planning (BBR) in Bonn, Germany. He has been in charge of the European Spatial Planning Observation Network EPSON, Contact Point Germany. Since 2007 he has been responsible for the domain Energy and Spatial Planning.

Josep Puig i Boix

Josep Puig is a consultant on energy and the environment. He works as specialist on renewable energy and the Spanish energy market with Ecofys. Also he teaches a course on Energy and Society in Environmental Science Studies at the Autonomous University of Barcelona. He has worked on renewable energy since 1981 in the University, with private companies and as a local politician. He is an energy engineer with a PhD and a Masters in Environmental Engineering. He has published many articles and books on energy and environment. He is vice-president of Eurosolar.

Viraj Puri

Viraj Puri is the Founder and Managing Director of Gotham Greens. Gotham Greens is creating New York City's first commercial-scale hydroponic farm. While a Project Manager for BrightFarm Systems, a New York-based environmental engineering and design firm, Mr Puri managed various urban agriculture projects. He has managed renewable energy and energy efficiency projects at the SECMOL Alternative Institute in Ladakh, India and at the Mount Mulanje Conservation Trust in Malawi, focusing on green building, passive solar design, solar photovoltaics and fuel-efficient cookstoves. He has delivered presentations on urban CEA at varied academic and private sector settings including Wageningen University, The Netherlands, Berlin Technical University and Ecocity World Summit 2008, San Francisco. He holds a BA from Colgate University and is a LEED-accredited professional

Anis Radzi

Anis Radzi graduated from the University of Sydney with a Bachelors degree in Architecture, summa cum laude, and a Masters in Urban Design. After practising architecture for six years she turned her attention to researching ways in which the built environment can become self-sustaining in energy terms, giving special focus to renewable energy integration in bioclimatic architecture and sustainable urban design strategies. Anis presently is on a doctorate-canditature track and involved in urban design teaching at bachelor and master levels.

Ariane Ruff

Ariane studied Geography at Potsdam University in Germany. From 2000 until 2003, she taught at the University of Erfurt, Germany. Since 2003, she has conducted a research project focusing on sustainable land management at Nordhausen University of Applied Sciences.

Hermann Scheer

Hermann Scheer studied Economics, Sociology, Political Science and Public Law between 1967 and 1972 at the University of Heidelberg and the Free University of Berlin. He received his PhD in Economic and Social Science in 1972. Dr Scheer was first elected member of the German parliament in 1980, and has been re-elected eight times since. From 1983 Hermann Scheer has been delegated by the German parliament to the Parliamentary Assembly of the Council of Europe, and served as Chairman of the Committee on Agriculture between 1994 and 1997. Dr Scheer has chaired as well as initiated numerous international research and development conferences. His work is dedicated to a broad shift in the energy basis of modern civilization from fossil and nuclear resources to renewable energies. He has demonstrated both the necessity and feasibility of this transition in his five books: *The Stored Sun* (1987), *The Solar Age* (1989), *Solar Strategy* (1993), *The Solar Economy* (1999) and *Climate Change. From the Fossil to the Solar Culture* (2002). In addition, Dr Scheer has also authored more than 1000 articles.

Matthias Schuler

Matthias Schuler is one of the managing directors of TRANSSOLAR Energietechnik in Stuttgart. Born in 1958, he was educated as a mechanical engineer at University Stuttgart. In 1992 he founded the company TRANSSOLAR Climate Engineering. TRANSSOLAR'S focus is on new energy saving and comfort optimizing strategies by an integral approach in building design. Today – with 40 employees in Stuttgart, Munich and New York – Matthias Schuler works on national and international projects with architects such as Kazuyo Sejima, Frank O. Gehry, Steven Holl, Jean Nouvel and Helmut Jahn. Since 2001, he has been teaching as a visiting professor at the Graduate School of Design, Harvard University, and he became Adjunct Professor for Environmental Technologies in 2008.

Michael Stöhr

Michael Stöhr, Physicist, born 1964, received his PhD in fundamental research on silicon at the European High Magnetic Field Facility in Grenoble, France. From 1992 to 2000, he worked as scientific collaborator on renewable energies at WIP, Munich. Since 2000, he has been a senior consultant at BAUM. Consult, Munich, where he coordinates international research projects on renewable energies. For the guidebook *On the Way to the 100% Region*, co-authored with his colleagues M. Tischer, M. Lurz and L. Karg from BAUM Consult GmbH, Munich, he has been awarded the German Solar Prize in 2006 in the media category.

The Stöhr family was Germany's Energy Saving Champion in 2006 in the category of tenants for having completely switched to fully renewable energy supply in its daily life. From 2004 to date, Michael Stöhr has been a member of the supervisory board of WOGENO München e.G., a co-operative for participatory, social and ecological dwelling in Munich, and has been involved in the planning and realization of the low-energy building where his family lives. In Munich's Messestadt-Riem quarter he has implemented two participatory PV plants.

Martin Vosseler

Dr Vosseler is an environmental activist who, as a physician, is motivated by his deep commitment to the health and well-being of people everywhere. Born in Basel in 1948, he studied and practised medicine in Basel (1982–1995) and worked as Research Fellow at the Division of Primary Care and Family Medicine of Harvard Medical School. Dr Vosseler is initiator and co-founder of PSR/IPPNW Switzerland (1981, Physicians for Social Responsibility, Swiss chapter of the International Physicians for the Prevention of Nuclear War, receiving the Nobel Peace Price in 1985): www.ippnw.ch; Physicians Action 'Air is Life' (1985): www.aefu.ch; SUNswitzerland (1997); International Energy Forum sun21 (1997): www.sun21.ch; and is a co-founder of Ecocity Basel (1986). He is the co-founder of Transatlantic21, 2006, with a Guinness Book of Records entry in 2007. Long distance walks are 1999 Konstanz, Germany to Santiago de Compostela, Spain; 2003 Basel to Bethlehem; and SUNwalk 2008: Los Angeles to Boston, USA. Martin Vosseler lives in Elm/Gl, Switzerland.

Andrew Went

Andrew received an Honours degree in Nanotechnology from Curtin University. He also has a graduate certificate in Sustainability Studies from Curtin University Sustainability Policy Institute (CUSP). He is now pursuing a PhD into the technologies required for V2G to become a reality.

List of Acronyms
and Abbreviations

AC	alternating current
ADFEC	Abu Dhabi Future Energy Company
BBR	German Federal Agency for Construction and Spatial Planning (Bundesamt für Bauwesen und Raumordnung)
BBSR	German Federal Institute for Building, City and Spatial Research (Bundesinstitut für Bau-, Stadt- und Raumforschung)
BedZed	Beddington Zero Emissions Development
BEV	battery electric vehicle
BMELV	German Ministry of Food Agriculture and Consumer Protection
BMU	Germany's Federal Ministry for Environment, Nature Conservation and Nuclear Safety
BTU	British thermal unit
°C	degrees Celsius
CATE	Cooperative for Technological and Energy Autonomy
CCP	Cities for Climate Protection
CCS	carbon capture and storage
CDTI	Centro para el Desarrollo Tecnológico e Industrial
CEA	controlled environment agriculture
CENER	National Renewables Centre
CENIFER	Integrated National Center for Training in Renewables
CEOE	Confederación Española de Organizaciones Empresariales
CF	capacity factor
CHP	combined heat and power
CNG	compressed natural gas
CO_2	carbon dioxide
COP	Conference of the Parties
CSA	community-supported agriculture
CSP	concentrating solar power
CSR	corporate social responsibility
CTE	Spanish Building Technical Code
DC	direct current
deENET GmbH	Competence Network for Decentralized Energy Technologies
DG TREN	Directorate-General for Energy and Transport
DH	district heating

DLR	German Aerospace Center
DSF	double skin facade
DSR	demand side response
DTI	Department of Trade and Industry
EEE	European Centre for Renewable Energy
EEG	Renewable Energy Act (Erneuerbare-Energien-Gesetz)
EHN	Corporación Energía Hidroeléctrica de Navarre
EMAS	Eco-Management and Audit Scheme
ESCO	environmental service company
ET	emissions trading
EU	European Union
EURATOM	European Atomic Energy Community
EWEA	European Wind Energy Association
EWG	Energy Watch Group
ExWoSt	Experimental Housing and Urban Development
FCAS	frequency control ancillary services
FEE	Force Énergétique par les Enfants
FiT	feed-in tariff
FNR	German Agency of Renewable Resources
GDP	gross domestic product
GW	gigawatt
HCT	hydrothermal carbon technology
HEV	hybrid electric vehicle
HVAC	heating, ventilation and air-conditioning
IAEA	International Atomic Energy Agency
ICE	internal combustion engine
ICLEI	International Council on Local Environmental Initiative
ICV	internal combustion vehicle
IDAE	Spanish Energy Agency
IEA	International Energy Agency
IECP	Integrated Energy and Climate Programme
IFC	International Finance Corporation
IPCC	Intergovernmental Panel on Climate Change
IPPNW	International Physicians for the Prevention of Nuclear War
IRENA	International Renewable Energy Agency
ITC	Technical Institute of the Canary Islands
IZNE	Interdisciplinary Centre for Sustainable Development
Kg	kilogram
km	kilometre
kt	kiloton
kV	kilovolt
kWh	kilowatt hour
LPG	liquefied petroleum gas
LTI	Long-Term Integration of Renewable Energies into the European Energy System

mb/d	million barrels per day
MDG	Millennium Development Goal
mtoe	million ton oil equivalent
MW	megawatt
NATTA	Network for Alternative Technology and Technology Assessment
NEM	National Energy Market
NEV	neighbourhood electric vehicle
NGO	non-government organization
NO_x	nitrogen oxide
NREL	National Renewable Energy Laboratory
OEM	original equipment manufactured
OPEC	Organization of Petroleum Exporting Countries
PER	Plan de Energia Renovable
PG&E	Pacific Gas & Electric
PHEV	plug-in hybrid electric vehicle
PIOH	Island Planning Regulations
PJ	petajoule
ppm	parts per million
PV	photovoltaics
R&D	research and development
REC	renewable energy certificate
REDP	Renewable Energy Development Project
SAFA	Finnish Association of Architects
SEEG	South Styria Cooperative for Energy and Protein Production
SEU	Sustainable Energy Utility
SMUD	Sacramento Municipal Utility District
SO_2	sulphur dioxide
SWF	Shaanxi Provincial Women's Federation
TARA	Tecnologías Alternativas Radicales y Autogestionadas
Tekes	National Technology Agency of Finland
TOU	time-of-use
TRANS-CSP	Trans-Mediterranean Interconnection for Concentrating Solar Power
TW	terawatt
UBA	German Federal Environment Agency
ULP	unleaded petrol
UNESCO	United Nations Educational, Scientific and Cultural Organization
UNFCCC	United Nations Framework Convention on Climate Change
UNISEO	United Nations International Sustainable Energy Association

UNSEGED	United Nations Solar Energy Group on Environment and Development
UNU-GTP	United Nations University Geothermal Training Program
US	United States of America
V2G	vehicle to grid
VAT	value added tax
VIG	vertically integrated greenhouse
WEO	World Energy Outlook
WTW	well to wheel
WWF	World Wide Fund for Nature
ZEB	zero-energy building
ZED	zero-energy district

Chapter One
100% Renewable Energy: The Essential Target

Peter Droege

Renewable power – foundation for human evolution

The time has come to abolish the combustion of coal, oil and gas for energy generation worldwide, along with the nuclear power threat. This is the historical challenge of today. It is unlike any other that preceded it in the emergence of human civilization. It is unprecedented because it involves a collective choice to be made across a wide array of technological, social and economic conditions – to go beyond the slow and messy process of blind evolution. A global renewable energy base is the very foundation of sustainable life on this planet. Only with it, massive afforestation efforts and lifestyle changes to higher quality and dramatically lowered material consumption become the essential elements of hope. A worldwide move to sustainable economic practice beyond green lip service may just still carry this promise: to rebuild the inherited system of wasteful abundance for the few into a basis for sustaining human life in a steady-state economy for all.

'100 per cent renewable' means an entirely renewable power base for the global economy, across the lifecycle of energy flows, embodied, operational, transport or stationary. In this world steeped in expensive and toxic hydrocarbon fuels and products it does not seem easy for anyone but isolated indigenous tribes to live up to this ideal. Nevertheless, the aim to rely on the abundant and largely free sources of the sun is clear, and it is necessary. A wave of innovations rises in infrastructure systems, personal transport or community development, successfully procuring non-polluting local electricity and thermal resources. Manufacturers begin to develop renewable production processes, and increasingly, producers of closed-cycle materials are keen on eliminating fossil carbon combustion content. This book is a snapshot of a dynamic picture, a world well on the path to sustaining human civilization on a renewable planet.

Is 100 per cent too ambitious? Climate change and fossil fuel production risks are now so massive that an anthropogenic carbon emissions balance has to be aimed at that is significantly below zero: current atmospheric carbon dioxide (CO_2) concentrations will have to be lowered by at least 25 per cent through carbon sequestration in forests and soils to eventually return to pre-industrial levels – that is, if a choice is to be made to actively and purposefully compensate for human damage. Others may prefer to pray instead for 'helpful disasters' such as, say, the collapse of the Gulf Stream to slow Greenland glacial melting.

How to get there? The path is different for each person, community, company or country. For some, internal renewable resources can be maximized more easily, for others, regional and national programmes will have to be the more powerful agenda carriers. Weak local government will require strong action by state and national institutions. Developing countries in the grip of international lending leveraged policies will also benefit from a reform of these policies, to advance 100 per cent renewable targets not merely as desirable aspects of sustainable development, but the very condition on which to found sustainable aims such as, for example, the Millennium Development Goals (MDGs).

Notes to a renewable future

After more than a century of fossil-fuel charged hyperdevelopment, the hydrocarbon age reaches its end. Predicted for half a century (Hubbert, 1956), conventional petroleum production capacity at stable prices has peaked, roughly in 2005–2006. As if to demonstrate this fact, along with its early consequences, an under-controlled United States (US) mortgage industry leveraged on highly fossil-fuel exposed land development – suburban sprawl – helped precipitate the great financial crisis commencing two years later. Economic growth under the conventional energy and resource consumption model has long become an oxymoronic notion (Meadows et al, 2004; Newman at al, 2008; Schindler and Zittel, 2008; see also www.abc.net.au/rn/ockhamsrazor/stories/2008/2445159.htm and www.resilientcitiesbook.org/index.php?page_id=281).

Meanwhile, greenhouse gas concentrations in the Earth's atmosphere exceeded proven sustainable levels of 280 parts per million (ppm) by at least one third – concentrations reaching 390 ppm in 2009 (Rahmstorf and Schellnhuber, 2007; Hansen et al, 2008; NOAA, 2009). Yet international climate negotiators and most other policy voices engaged in international climate negotiations misinterpreted Intergovernmental Panel on Climate Change (IPCC) advice, declaring 450 or even 550 ppm CO_2 in atmospheric concentrations as safe – as if the solemn incantation of an artificial number approaching twice the pre-industrial levels would be adhered to by the biosphere. Science and politics merged, even as planetary systems raised the spectre of out-of-control climate destabilization at 380 ppm and a 0.7 degrees celsius (°C) temperature rise. Saner voices call for 280 ppm or at most 350 ppm (Hansen et al, 2008), even if taking a rather principled stance – a target repeated by Al Gore at the climate talks in Poznan, December 2008. Gore spoke to the great cheer of the audience, while the main conference agenda was still focused on avoiding any targets – or accepting the magic 450 ppm at best.

This book focuses on the only desirable consequence of this historical stage in humankind's evolution. Its authors document key elements and features of a complete turn away from fossil fuels and nuclear power. They and their projects are living testimony to the extraordinarily practical, exceptionally everyday and stunningly normal transition from the mechanical age of oversupply by rigid 'base load' – a characteristic of the power generating dinosaurs of the early industrial age – to the vision of a planetary response more in harmony with the diurnal rhythm (Mills and Morgan, 2008) and geographic conditions.

With this shift the seasons begin to matter again, the difference between night and day, local weather patterns and the specific advantages of location, including cultural characteristics. None of this seemed relevant in the dark days of the conventional power regime when maximizing output was the call of duty and reward – discounting the planetary pollution with toxic gases, carcinogenic particulates, lethal radioactive material, and the dominant culture of wasteful abundance that marked the 20th century's industrialized regime of consumption. In a renewably supported world, so we hope, humanity can begin to breathe and prosper, freed from the most powerful shackles: the conventional energy chains.

Figure 1.1 Two oil wells burning side by side at Santa Fe Springs, California oil field, 1928

Source: Los Angeles Times, 19 September 1928,
Los Angeles Times photographic archive, UCLA Library

Raising a phoenix from the planet of waste

Fossil fuel burning has been the main source of a pernicious accumulation of airborne industrial waste, followed by deforestation and industrialized agriculture. Yet this worldwide incendiary frenzy will cease eventually, with strategies implemented to remove the human-induced excess greenhouse gas from the atmosphere. Renewable energy democracy, founded on the responsible footing of lowered demand and greater efficiency, is the only logical heir to the conventional power dictatorship. Unlike the old system, it is based on ubiquitous sources and capable of operating at a local, community, national and global scale. To put it differently: to work in sustainable ways, and to assist without squandering another precious moment, this virtually limitless source of energy must be deployed not in an attempt to simply replicate the toxic sources inherited, but with an entirely new way of looking at energy generation, conversion, trade and use. This also implies bold reductions in primary energy and resource demand, greater efficiency and more just and equal ways of approaching international relations and global development. Great changes lie ahead.

Globally, outdated energy systems have reached a crisis point, given notorious failures to respond to new realities and to plan and act in the face of a long-developing crisis. The outlook of an exhausting resource base has been all too clear for more than a generation, and expressed in an overwhelming reality of pollution and poverty around the globe. A clairvoyant and daring glance into the dark pit of an unmitigated petroleum supply slide reveals air travel, shipping and ground transport as severely curtailed, some cities that shrink and others that fail, and oil-denominated financial systems collapsing. The very relation of nations, of people and community life is affected; the thin fabric of civility fraying in several regions of the world, amidst the scurrying for remnant oil reserves, or the anticipation of migrating climate refugees. Such spectres can still be reduced or even avoided if unprecedented steps are taken to modify, shift, transform and replace the present system, while preparing societies and nation states for the great energy and climate transformations to come.

From an age of fossil-fuelled mechanization to a responsive era of renewable autonomy

Fossil fuel-charged industrialization, first the electric and then the eclectic electronic revolution have compounded the complexity of human civilization and increased its interdependence. A legacy of globalized and tightly interconnected resource flows resulted but also led to great social imbalances evident in vast pockets of poverty around the world. A mechanized, homogenized and corporatized global food production, distribution and profit system has emerged with it, epitomizing the current crisis of resilience. In this centralized and excessively interdependent environment, both the primary export producing and the industrialized worlds become less resilient to supply

Figure 1.2 Power lines into a brightly lit Sheffield and the Meadowhall Shopping Centre over the M1 motorway, with the disused Tinsley cooling towers: The era of the discardable civilization has come and gone, and its infrastructure of waste has become an embarrassment to many countries

Source: Alan Hood

Figure 1.3 Windmill in Crete, Greece: The return of appropriate technology – the new technology is advanced but the principles are timeless. Good systems are still those appropriate to location, season and societal setting

Source: Martin Carter

shocks (Homer-Dixon, 2009). The fossil-fuelled world is a world of tragic irony. All people struggle for progress and a better life, driven by hope for survival or more wealth, but in basing these hopes on false assumptions of a failing energy and resource consumption model, all work towards decreasing stability, lowered levels of security and a quickly eroding natural resource base.

Renewable energy exemplifies an entirely different paradigm, that of decentralization, local resource reliance and regional autonomy. This is no Luddite or isolationist vision, but a necessary move to locally and regionally self-sufficient models of development. The global trade in oil, coal, gas and uranium has nothing to contribute to sustainable forms of prosperity. Here the local and regional conversion of solar power, wind, water, subterranean heat sources and bioenergy, combined with reformed soil management and reafforestation, offer the key solutions for a new age – a certain degree of simplification by shifting complexity from the global to the regional and local realm.

Facing a conversion investment challenge

If the slow progress of the past is any indication, it is essential to divest radically, not gradually, from old energy systems and outmoded building and infrastructure design approaches. There is a larger economic reason for this and it lies in the conversion investment challenge. After discounting inertia, vested interests and attempts to hold onto past glories and fading riches, what emerges is that, in principle, equally significant resources are required to invest in new or in old infrastructure. Funds that continue to be expended on antiquated systems represent wasteful investment in failing infrastructure and directly compete with much needed resources for new technology. This also diminishes the reserves needed to build a new and survivable existential base. More troubling is the still ill-understood fact that civilization is in overshoot mode. The most significant challenge of economic policy today is to engineer a soft landing into a stable state (Meadows et al, 2004).

The resources of strong times must be spent on averting the times of need that otherwise are certain to come. The final years of the old energy world must be used to convert investment to the new infrastructure – at a rate and with a focus that far exceeds what is being done today. To move to a fully renewable infrastructure is the most pressing task for global diplomacy, policy and social reform agendas.

What is the effect of the 100% renewable target?

'100 per cent renewable' encourages governments, businesses, cities and communities to leave behind programmes that only superficially 'green' development strategies, deploy the same production methods and material flows while embellishing them with the odd efficiency contract and one or two energy-related corporate social responsibility (CSR) pledges, perhaps even some solar roof elements or the occasional co-generation plant. It signals to the automotive industry that this is not a matter of taking a business-as-usual fleet of the same old cars, but of shrinking them, lightening them, networking and endowing them with electric hearts and electronic brains – and making them available on short- and long-term leasing bases. It entails a shift from rote car manufacturing, with success measured in sheer numbers, to a business model structured around a range of advanced commercial and consumer mobility services. This would mean a massive shift in ownership arrangements, vehicle use patterns and manufacturing techniques, dramatically lowering energy and primary material requirements.

For the individual household, 100 per cent renewable, in its full consequence, also means depending on a fully renewable resource stream: being able to rely on, say, total fossil energy content that is lowered by 80 per cent when compared to conventional practice, when looking at the total balance of goods and services consumed, with shortfalls compensated by renewable power. It means relocalization and re-regionalization of energy systems, with radical care and surgical attention spent on the unflinching severing of what Hermann Scheer has called the long fossil energy chains in his *Solar World Economy* (Scheer, 2002). The successful, coordinated decapitation of the fossil hydrocarbon Hydra is necessary to rescue the world economy from the long, carbon dioxide belching and deeply impoverishing lines of the petroleum age, restructuring it to the short and locally empowering networks of the renewable era. To many this may still seem like a distant ideal, yet it is a practical and immediately necessary aim. It offers rich opportunities to most incumbent industries, waiting to be charged with new life and new meaning, in the vast and ubiquitous world of renewable energy conversion, storage, dispatch, operations, investment and management services.

The seemingly Herculean 100 per cent target is the very rationale behind this book, prompting an important disclaimer. This book contains no examples or references that fully live up to this ideal – yet all of its authors contribute and share important plans and visions – and many demonstrate practical steps on the path to achieving these. And another caveat is offered here: easy claims of being 'fully renewable' or the even more dubious, 'carbon-neutral', have become rationales for a narrowly conceived, conventional practice. The environmental benefits of renewable energies have given rise to a deluge of green-washing and less-than-beneficial initiatives, continued over-consumption and prolific, mutant green business-as-usual.

100%: A target to end all targets

'100 per cent renewable' is a call to build an advanced civilization, rather than to nervously focus on futile fractional emissions targets. Climate- and energy-risk conscious communities and countries are galvanized by targets: 20 per cent by 2020, 50 per cent by 2030, 80 per cent by 2050. This statistical contest is understandable, and may have been healthy at some point. Yet the wrangling for targets is also misleading since it is far from clear what these percentages mean, given the many different ways of calculating and monitoring, let alone arriving at them. Many target pledges have served to defer or even substitute for action; and the more distant the target the less meaningful it becomes. To 'back-cast' a possible path from a distant goal and simply follow it is a deceptively simple idea in theory but does not work in practice, given the vagaries of chance and the discontinuities built into political processes. Many early target setters were either too optimistic, or not ambitious enough, or lacked clear ways of knowing how to define, determine, compare or enforce targets. After a generation of development, there is still no unified, universally accepted and practiced system of target and baseline setting, not for companies, urban communities, countries – or the world.

All too often in those comparably halcyon early days of the publicly acknowledged climate emergency, government planners in cities, states and multi-state unions, along with the chiefs of major energy and manufacturing corporations became occupied with the art and science of performance accounting, more so than with the design and execution of defossilizing action programmes, the pursuit of partnerships in material reincarnation and product longevity, the financing of innovative technology infrastructures or purposeful institutional reform. The dead weight of political expediencies and false economic dogmas aside, the reason for this agonizing delay has been a gross and systematic underestimation of renewable capabilities. The potential for a far more rapid and deeper change than imagined was suppressed or ignored, given that the policy conditions required for efficiency and renewable generation were seen to threaten established interests. To be taken seriously, today's targets have to be radical, measurable and pursued with unwaivering commitment, with all necessary resources and accountable commitments in place. As the sense of urgency increases, distant fractional targets become increasingly meaningless.

Yes, we can – and we must

According to some sources the global petroleum production peak has passed and its rate of decline is expected to exceed 3 per cent annually (Schindler and Zittel, 2007). The atmosphere is over-saturated with gases trapping sun radiation. The acidifying, warming oceans have begun to reject rather than absorb CO_2. The call for 100 per cent is therefore not a utopian polemic against fossil fuel use, social injustice and over-consumption, but a sober call for securing the continuity of civilization. In pushing the replacement of fossil fuel combusting systems with the power of the sun, the wind, the earth and the land, the only sensible and ultimately comparable target is not 10, 20 or 60 per cent by such-and-such a date, but no less than 100 per cent, to be reached as quickly as possible, by the most effective means available.

But is '100 per cent renewable' really achievable? Depending on whether primary or final energy is calculated, at the beginning of the 21st century only between 12 and 20 per cent of all energy commercially derived from all sources was non-hydrocarbon based. And only some three to four-fifths of this was in the renewable category, according to serious world energy statistics (USGS, 2005). In a world so overwhelmed by oil and choking on coal, it is difficult – to put it mildly – to construct a renewable life in any pure sense. The dominant fossil fuel reality determines individual lives and community practice in so many ways. Most aspects of even the most peaceful, serene civic lives, from food to pharmaceuticals, are linked to oil fields and coal mines, and implicated in the epic and brutal oil wars from Iraq to the Sudan, and the environmental and human toll wreaked by coal mining.

The link from distant resource atrocities to every individual human life is established through the consumption of goods and services, and the use of the US dollar and other international currencies that are based on fossil fuel as the new gold standard. Indeed, to some, one major success indicator may well be the prevalence of regional 'solar dollars' (proposed here as the *sollar*) – stable currencies not based on conventional interest rates and tied to the time-limited barter value assigned to regional goods and services, but linked in exchange worth to, say, the value of one kilowatt hour of solar energy.

Reduced demand and greater efficiency aims are important but not sufficient

Today, action comes on various fronts. Relevant studies confirm that a demand reduction and efficiency gain of at least 50 per cent – and in especially wasteful examples, often 80 or even 90 per cent – across all sectors is technically achievable and an essential base on which to found a renewable future. A recent European Renewable Energy Council study sees the potential for a full 75 per cent of global carbon emissions reductions to come from efficiency measures, against those projected under the business-as-usual scenario (EREC, 2007).

Beyond this, a massive phasing in of renewables is needed, particularly since efficiency improvements can be slow, or have in some countries even gone into reverse. They do not nearly keep pace with increasing demand; worse, efficiency savings are notoriously taken up by a rise in use. And beyond efficiency and renewable generation, dramatic lifestyle changes are needed. Household consumption of goods and services make up a lion's share of total energy use. Yet very few local energy statistics reflect this fact, and fewer still are the local government practices and programmes that embrace a shift in consumption patterns (Lenzen et al. 2008). The World Wide Fund for Nature (WWF)-sponsored One Planet Living concept and its Beddington Zero Emissions Development (BedZed) project are still rarities. This commercial package was launched as a local, community-grown initiative, and while it had intrinsic model aims it has so far failed to be broadly influential. Today WWF attempts other, more visible, grander partnerships, for example, hoping for the limelight in which Abu Dhabi's Masdar City basks. The message is to be applauded, if it indeed means 'one-planet living': realizing an existence within the means allotted to each human being, measured on a global equity basis.

Figure 1.4 The environmental footprint of an average American or Australian family: The average American or Australian suburban family of four, residing on a 600 metre square lot, requires 56 hectares in land for food, energy, products, services and waste assimilation – not even counting the area required for sequestering carbon emissions. In order to not exceed its global fair share, it will have to reduce its environmental footprint by 86 per cent, to some 8 hectares

Source: Image courtesy of Richard Weller, Donna Broun and Kieran McKernan; data from Johnson (2006)

Expanded to the current state of affairs, one-planet living would mean the immediate elimination of petroleum-nurtured, long-haul beef from global diets – and of much other animal protein. And without a restructuring of global trade relations away from the wasteful, simple-minded and grossly inequitable export–import divide of the world, to help rebuild local self-sufficiency and sustainable export capacity, it will be near impossible to swing around the virtual Exxon Valdez that is the current development model.

To recapitulate: few people, communities or companies in the industrial world, and certainly no country or city, pursue lifestyle change beyond gestures and lip service. Yet seen on a per-household basis, energy and resources embodied in lifestyle choices often outstrip direct energy use in many modern cities, including that of transport. And efficiency is not pursued as seriously as is both possible and needed. Instead, the word 'efficiency' is recited in endless incantations as a kind of absolution chant: to do so is seen as safe practice since it does not question the fundamental carbon and uranium base of industrial energy supply. At the same time conventional power supply systems have been over-designed to maximize fuel consumption, and based on the brute force of wasteful base load power systems scaled to match peak demand – making a mockery of efficiency programmes.

Indeed, in this paradoxical world, fractional carbon emission reduction targets ridicule individual energy saving or solar installation efforts: any gains above the target are taken up by reduced efforts on the part of commercial polluters. And the very notion of carbon trading as an overriding policy maxim has introduced another paradox with its own corollary loophole: an investment in a relatively minor efficiency improvement to a coal fired power plant, while cementing its CO_2-belching existence for another 30 years, is treated as equal to attempts at replacing the carbon-based paradigm altogether – as long as the emission reductions add up in a comparable manner. And here the powerful incumbent interests win out. The source of the emissions, not the emissions themselves are infinitely more useful a focus in any serious target setting exercise.

Pursuing '100% cent renewable' as a practical goal

This book is a window on practice operating in the vicinity of the 100 per cent renewable ideal. It is about a number of approaches, more and less partial, none perfect, some more, some less flawed. But jointly, these references and models are a powerful encouragement to embrace 100 per cent as the only sensible target to be pursued at this very late point in time.

One way for a company, community or person to reach a theoretically and practically pure, 100 per cent renewable support base is to escape the carbon spider's web by living in isolation, on a real, virtual or artificially defined island. Indeed, the construction of partially energy autonomous islands of sorts is the secret of success for many fully renewable entrepreneurs at all levels today, from Surrey's Woking Council to Israel's Better Place. To the hopeful and optimistic, these islands will proliferate fast and extensively as they connect into a new and renewable reality.

But the label '100 per cent renewable' in the somewhat less-than-ideal sense can also apply to aspects of people's lives, or communities and their support structures. For example, households and businesses can operate their own domestic or corporate renewable energy system by installing photovoltaic panels, a woodchip-fired combined-cycle heating plant or biogas powered tri, quad- or even quint-generation system, also treating and recycling water and sequestering CO_2, besides supplying power, heat and cooling. Or local governments can act in this way by investing in a solar thermal field or a wind farm, or by contracting the delivery of renewable electricity from a private supplier, within or beyond the local borders. For instance, a new partnership in solar power development will help yield 100 per cent renewable electricity for all households for a large city such as Munich, Germany – a city long devoted to gains in energy efficiency in its building stock and district heating networks.

As another example of a partial, or 'systems focused' 100 per cent policy, the UK government pursues the goal for all new homes to be 'zero carbon' by 2016 – implying the need for a massive proliferation of local, distributed energy supplies. The fact that only 0.1 per cent of UK renewable energy capacity is generated on public land has been used to illustrate both the woefully inadequate state of local community and city participation in renewable generation and its extraordinary potential as a future source (Slavin, 2007).

The right policy matters

It is also important to distinguish the quality of policies by their ability to deliver a fully renewable world. What policies are capable of achieving or supporting the development of 100 per cent renewable 'islands', infrastructure elements or behaviour, and which are less promising? Emissions trading (ET), for example, seems like a compelling theory to many market economists. In theory, ET schemes can help convert pollution debit-derived income into desperately needed investment in carbon sequestering soils management, organic agriculture, biochar/pyrolysis programmes and general, massive forestation programmes, and successful efforts at permanently lowering fossil energy-based generation.

Today, most ET programmes are used as a general revenue raising tool for a wide range of purposes – not for narrowly specified beneficiary projects. Devised such, ET is unable to accomplish significant improvements effectively, and without being introduced as transitional sunset schemes, as in auctioned permit-based cap-and-trade arrangements, it cannot lead to the phasing out of fossil fuel use, just like emissions offsets cannot logically be an effective means of reliably reducing emissions. As currently practised, ET can and does marginally support some renewable energy programmes – but at the intolerable cost of extending the life of the antiquated systems of non-renewable power generation. Without a definite zero horizon and vaguely defined beneficiary projects it does its work incidentally and blindly, following a theoretical logic of resources flowing to the most efficient, lowest-cost means of delivery, while in reality most credits flow too often to the most powerful polluters in the industrial world, as perverse windfall profits. Examples include the conventional power supply industry under the disastrous early European experience with ET, and the notoriously inefficient Clean Development Mechanism. Undeserved profits were also reaped by the early purveyors of the globally unworkable carbon capture and storage projects, or offset schemes with uncertain, even net-negative benefit (Lohman, 2007). And in a world of transnational corporations based trade polluting industries risk being merely shifted to a pollution credit-rich developing world – the 'non-abatement countries' in UNFCCC parlance, supporting the proliferation of high-emission practices there (Schreuder 2009).

ET as a concept has been transferred from one context to another, without considering important distinctions that make this transfer unlikely to succeed. Derived from the sulphur pollution trading experience in the US, from a national provision for a relatively reversible pollution agent with regional impact without terminal time restrictions, it is now applied to a global, not at all easily reversible problem with global impact and an extremely critical time

Figure 1.5 Not 100 per cent

Source: Ashden Awards/Renewable Devices Swift Turbines

limitation – so critical that time may in fact already be up. To function on the debit side as an effective emergency measure and transitional market arrangement, carbon trade would have to be comprehensive and result in the dramatic and timely reduction of greenhouse emissions. On the credit side, it also would have to be limited to key, proven and broadly, immediately effective applications: the storing of carbon in soils and the generation of pure and new renewable energy, and therefore not cleaner gas, petroleum or coal combustion technologies, carbon emissions sequestration or nuclear power.

To reiterate: to be effective and useful ET must efficiently function at this high level of focus and administration on both debit and credit aspects. But neither is the case: the sulphur trading experience does not lend itself to application on a global level in any way; and due to its many complexities, ET has been used to slow, not speed up progress.

Attempts to apply carbon debits and credits to individual efficiency or even urban-scale projects are bound to get mired in accounting, certification and monitoring – massive costs arise. By counting proximate, inefficient uses and users of energy, indirect responsibilities are identified and double-counting occurs. Because of the nature of power politics, costs are shifted from primary polluters to consumers. Effective ET also requires a firm time horizon, a sunset clause aimed at the ultimate elimination by conversion of fossil energy sources. A healthy example is the early Obama administration's commitment to a 83 per cent reduction in carbon emissions by 2050, and the linking of a cap-and-trade scheme with auctioned permits to this – except the final target should be at least 95 per cent, to meet basic climate stability aims, given the present state of scientific insight.

Sloppily propagated, misapplied and weakly administered as it is today, the carbon trade perversely depends on a strong fossil fuel regime to provide the funding for so-called 'low-carbon projects', via pollution permit payments – without delivering the benefits anywhere nearly as efficiently as other means. Therefore, many argue with US economist Gilbert Metcalf that if emissions are to be relied on as a policy tool, rather than a more direct focus on their root cause as fossil fuel combustion, a carbon tax aimed at producers of coal and oil would be a more equitable policy-focused and effective approach, raising US$85 billion annually on a $15 per ton carbon levy (Rotman, 2009), but carbon tax is used in too few places in the world today or not even discussed much, thus having little chance of taking the place of ET. To some it still carries the unhappy connotation of, well, a tax. And it, too, builds dependence on the revenue from polluters, is difficult to administer fairly, and in itself is too blunt a weapon in the battle for achieving serious reduction outcomes, particularly since Metcalf and others propose to spend the revenue on personal income tax credits, not primarily or directly on renewable energy infrastructure. The carbon tax revenue is here used as a simple pricing mechanism, to notionally internalize some environmental costs, with added costs passed on to the consumer – who then receives the tax increase back as a credit. Used as a circular scheme the revenue does not promise to be effective in curtailing emissions nor fostering new technology – nor is it available to innovators, cities, communities and others that hope to not only benefit from fair environmental pricing that effectively approximates external costs but also derive reliable support for individual renewable power production. One challenge in attempting to internalize the external costs of climate change is that the price may have to be infinite – given the potential size of the damage.

Renewable energy certificate (REC) trading was intended to help create a fairer marketplace, but this mechanism, too, has proven to be a disappointing, limited policy tool. This is unsurprising not only because REC trade, like the carbon exchange, artificially operates within a global bubble of fossil fuel subsidies, but also because it gives rise to a complicated and cumbersome system not easily accessible to small and new players. Also there are risks such as those arising from uncertainty in future certificate values and the wholesale price of electricity, which raise investment costs significantly. For RECs to have a value they require mandatory renewable energy targets to be set. Minimum targets can become maxima, or can be so quickly taken up by a relative excess of existing renewable generation assets that an REC market can collapse. A fundamental policy flaw has been to artificially limit a notional future target, or to set too short horizons.

But there are other, more effective tools. These also work by framing markets, markets that are supported by just policies such as national or state feed-in tariffs (FiT), off-grid, thermal-energy and other targeted regulatory frameworks, enhanced by revenue or credit support signals and renewable-content pricing, all supported by the dramatic reduction or elimination of fossil fuel subsidies. Governments, including local authorities, can also institute development, production or acquisition support incentives for renewable technologies. Examples include support for electric car concepts that are connected to renewable energy supply systems, through zero or strongly reduced import, sales or excise taxes, loans, grants or even, as a transitional measure, fee-bate schemes.

Feed-in tariffs

Some schemes are more efficiently focused on renewable electricity introduction than others. The FiT, described in Chapter 7, has been shown to be the most efficiently focused, effective and equitable way so far of not only achieving rapid greenhouse gas abatements but also producing renewable electricity at the lowest rate of all policy frameworks available (BMU, 2007). They represent one of two public policy mechanisms available that are capable of triggering and guiding a 100 per cent renewable conversion. Misinformation campaigns created the impression that FiTs are expensive, but the opposite is true. The German Federal Environment Ministry reports that during 2007, the FiT cost a mere EU€35 per household (BMU, 2008), while it proved drama-tically effective in generating employment, delivering innovation impulses to research and development (R&D) and manufacturing, and mitigating fossil fuel price rises. In that year Germany installed more renewable capacity than the UK managed in a decade with its certificate-based system, lifting its share of total electricity to 8.5 per cent, from 7.5 per cent the previous year. Together with efficiency improvements, a real savings gain of €5 billion is projected for 2020, when compared to the business-as-usual scenario. Other mechanisms include the removal of all fossil fuel subsidies, direct, indirect and hidden, and design pricing mechanisms that allow all external costs to be reflected: from the massive public health cost incurred by fossil air pollution to the cost of oil-related wars in the Middle East and Africa.

The 100% target faces no technological barrier

A sufficient number of theoretical studies and model simulations has been conducted to demonstrate that virtually all countries are capable of attaining a fully renewable energy electricity and wider energy supply system, using a range of technologies (ALTER, 1996; Lehmann, 2003; Peter and Lehmann, 2004; Trieb et al, 2006). These models and empirical investigations indicate that substantial benefits are to be expected from lowered costs in health, ecological damage and energy expenditures, to employment gains and lowered military risks. Giving further credence to the practicality of the 100 per cent goal are many cases illustrating how the market, once even only slightly relieved from its fossil and nuclear shackles, richly rewards investments in renewable choices with an enormous stream of innovations. Until the turn of the century, an arduous

Figure 1.6 Solar energy installation in Weizmann Institute, Rehovot, Israel

Source: Alla Leitus

path seemed to stretch from contemporary theory and minority successes to global, mainstream practice. This outlook very much changed in the 2000s – certainly in the areas of large-scale wind and sun power installations, as feed-in tariffs and tax incentives came into effect, and electricity grid price parity between coal and solar energy approached in large markets such as the US.

Renewable energy investments multiplied, and installed capacity skyrocketed, admittedly from the very low base where it had languished at for years. The potential is enormous. Stanford University in the US was the home to a study modelling global wind power capacity alone, suggesting that all commercial energy demand, including transport, and more than seven times global electricity demand could be satisfied through wind energy alone, if only 20 per cent of global potential locations with a 6.9 metres per second wind speed at 80 metres above ground were provided with standard 1.5 megawatt (MW) turbines (Archer and Jacobson, 2005). The evidence had long pointed in this direction, with Germany and other European countries beginning to reconfigure their national grids to anticipate the enormous if mildly stochastic power to be derived from the burgeoning wind fields.

David Mills, Canadian-born physicist long based at the University of Sydney, provides another example of the capacity of large or municipal-scale systems. Rescued by Silicon Valley entrepreneur Vinod Khosla's venture capital ingenuity, Mills escaped the mire of Australian tragically coal-power dominated policies and was embraced with open arms by California, the state governed by solar

Figure 1.7 Californian governor Schwarzenegger opening Ausra's Bakersfield pilot, 24 October 2008, the first new Californian solar-thermal plant constructed in 20 years

Source: Ausra

Figure 1.8 A cooling tower for a geothermal power plant in New Zealand

Source: Allan/morgueFile.com

supporter Arnold Schwarzenegger. In doing so, he joined the long tradition of municipal-scale solar energy initiatives in the country's southwest. As a physicist more comfortable with utility-scale technology than with local community solutions, he also abandoned his long-held hope for integrated energy systems in cities, realizing that urban communities require adequate municipal power institutions, complex arrangements and partnerships to change from within.

From his new company Ausra, Mills quickly began to conceive and implement solar thermal plants based on his linear Fresnel-lens based steam power plants, building a 25MW demonstration facility in Bakersfield, and planning a field seven times that size nearby. Mills' hopes are for the gigawatt (GW) and terawatt (TW) supply needed to replace coal and even petroleum for transport, worldwide (Mills and Morgan, 2008). Photovoltaic protagonists are equally confident: only some 2.5 per cent of the 250,000 square miles of available land in the US southwest are said to be required to match the country's total 2006 energy consumption, along with the high-voltage, direct current (DC) power lines required to ship the electricity to consumers, for only $420 billion in subsidies between 2011 and 2050 (Zweibel at al, 2008).

Geothermopolis

The escalating search for urban-scale clean power sources substituting for conventional sources raises the spectre of geothermal new towns rising – expanding cities and suburbs located near geological heat stores, much like early coal towns in the UK forming near mines, rather than in association with established urban areas. But this seems to be a minor issue compared with the many positive examples of existing cities expanding their geothermal sources, following veteran *geothermopolises* Reykjavik, Iceland, or Rotorua, New Zealand. These include Anaheim, California; Reno, Nevada; Boise, Idaho (Jordan, 2009; see also www.c40**cities**.org/bestpractices/renewables/reykjavik_ **geothermal**.jsp, www.c40cities.org/bestpractices/renewables/ reykjavik_ geothermal.jsp, www.cityofboise.org/Departments/Public_Works/ Services/ Geothermal/index.aspx, www.trendhunter.com/trends/paris-gets-green-heating, www.anaheim.net/utilities/news/article.asp?id=663, www.geoheat.oit. edu/bulletin/bull17-1/art4.pdf); or Munich, Germany, where a long-established relation with heat carrying aquifers affords virtually 100 per cent carbon-free heating in some urban redevelopment areas.

Summary: What are the challenges to be overcome?

Addressing challenges today means avoiding being disappointed tomorrow. There are a number of ways in which 100 per cent can be implemented without major structural challenges. Examples are the many smaller communities, islands and partial systems referred to here. On a larger scale, the challenges are more formidable but not insurmountable. Several broad issues arise at this level and we examine how to address them here.

Reforming institutional arrangements

Most modern governmental and civic institutions were shaped during, by and for the fossil fuel economy and its reign over lives, societies, countries and ideas about the global order. Its rule is epitomized in the great successes of the 20th century, but also in many of its failures. As the fossil era wanes and gives rise to a renewable world, established municipal, state and national – even international – organizational structures and response mechanisms are no longer appropriate. They were tailored to an age of centralized, one-way and oligarchic power supply structures, and a stark separation between civil and energy-industrial realms of decision-making. A central feature of this separation included the rampant privatization of municipal and other public utilities throughout the 20th century and the modernist maxim of maintaining a strict divide between mass consumers and producers.

Opportunities arise in broad institutional reform, the move towards outcome-oriented and accountable organizations, and a growing recognition that local political leaders and public servants can benefit from boldly leading a renewable energy geared path. The challenge here is for communities to not be satisfied with meeting the same old objectives through new management arrangements focused on process efficiency alone, but to make the transformation to a 100 per cent fossil fuel free community both the object and the measure of accountability.

Understanding vested interests and different interests

Global nuclear and fossil fuel lobbies represent an entrenched, powerful and well-funded force, smart and driven by an abiding mission to increase volume and market share. They know how to build allies, and learned how to look good in a smart green suit. These industrialists manage to turn handsome profits even from global warming. Billions in carbon credits have been claimed and banked, and windfall profits reaped by polluters in the industrial and developing world. These free pollution allowances, carbon credits and perversions of the system complicate change, are inflationary without yielding benefits and deprive renewable energy suppliers of precious resources. While there are many genuine innovators, 'going green' is the new black for oil companies, airlines, large developers and business-as-usual banks – yet all too often without shifting much more than the headlines on their annual reports. This can take the wind out of the sails of genuinely renewable initiatives, quite literally.

There is also an enormous gap in understanding or broad appreciation of the great differences between conventional, centrally sourced and controlled, and renewable, distributed energy systems. Fossil generators and nuclear reactors use increasingly rare, toxic and expensive resources, while most renewable sources are based on free, virtually unlimited and ubiquitous origins: sunlight. Associated with this difference are vastly different business models and interests. Power and profits of conventional miners, generators and distributors are inextricably linked to the value of the primary source and its processing and management risks and costs. Mining and processing industries are vast, and their interests well understood and expressed in many halls of political power. By definition, renewable industries are smaller in scale and fragmented, and the harvesting of sunshine, wind or water sources, and to a large extent also that of geothermal and bioenergy assets, is widely distributed across the globe – much unlike coal, petroleum and uranium deposits.

By resisting the inexorable end of the fossil power era, the dream of extending its life through pollution trade has also evoked a nightmare. Yet here lies an opportunity for the 100 per cent entrepreneur. A growing number of companies and their CSR unit heads, board members and investors grow weary of the lack of progress while lip service continues to be paid in rich quantities. They grow anxious about the lack of progress in combating the processes of global ecocide, and in capturing opportunities arising from the historical innovations prompted by a global technological shift. Above all, many champions of vested interests, too, realize that the greatest threat to future prosperity lies in rigidly holding on to assets of the past. These companies and individuals are the allies in the struggle to allow the greatest and most hopeful paradigm shift of our times to take place.

And the timing is good. While the fossil fuel, coal generation and nuclear energy complex has been built up over a century, and has an enormous network of finance, mining, distribution, power conversion and sales interests, with the policy and R&D capabilities wielding massive influence among decision-makers, this world is also highly dynamic and inexorably disintegrating. Smart players hedge their bets or look for exit strategies. Responsible leaders focus on the energy systems and services business and turn their backs on fossil and nuclear power altogether. And those with foresight and planning acumen have already embarked on fully renewable ventures.

Going beyond carbon trade

Emissions trading schemes have the potential to become the Trojan horses of climate policy. They look convincing and impressive from afar but contain an aggressive, messy and potentially fatal cargo: allowances for the worst polluters, the global migration of high-pollution industries to low emission countries, hilariously inappropriate rewards to dubious abatement schemes in many developing countries, a cumbersome and expensive process, and a logic that is fuzzy and vaguely geared towards 'low-carbon' schemes and actions. And by being paired with a blind posture towards a range of so-called low-carbon technologies and schemes they fail to distinguish the broad and long-term advantages of renewable power from poor *ersatz* schemes such as carbon capture and storage and other 'clean coal' technologies, notional deforestation plans temporarily foregone, carbon sequestration crop planted without proper planning or management, or biofuel projects in virgin forests. And they are at best too slow: visible evidence from the Arctic, Antarctic, glaciers, ocean acidity levels, biodiversity trends, and a host of corollary scientific observations suggests that by the time a global emissions trading regime may possibly be framed and its variegated controls and mechanisms reticulated to a regional and local level in an equitable, transparent and agreed manner – and if it could ever be made to work – it will be far too late to make a difference. Bold action has long been overdue, and to continue to bank on merely a round number, say, the magic 2050 target timeframe, or notional emissions targets or caps with substantial pollution rights granted, seems more like wishful thinking than sound public policy.

The opportunity here lies in working with but not only within the process, that is to say, not expecting locally workable solutions from a global concept, or relying on it as any form of salvation. While some wait for an impossible climate Camelot of perfect trade to zero emissions to emerge, the answer lies in implementing regional and local action plans as a matter of urgency. Successful 100 per cent renewable implementation champions understand the weaknesses and flaws of the Kyoto processes. And hence while they understand the sentiments and potential benefits, they do not wait with implementing 100% concepts, be they community or company-wide projects, that is, renewable energy island schemes or networked infrastructure systems, or both.

Nested supply: A new geography of power and a new social contract emerge

No single level of support will work on its own. At a country or continental level, comprehensive grids are being discussed and advanced to facilitate renewable energy interconnection. These massive connections can involve high-voltage DC links supporting long-distance transmissions at a low loss rate. Continental solar fields, biofuel resources, offshore wind parks and the like are envisioned to be linked with pumped-storage and other reserve systems to replace declining fossil and nuclear power resources. At a national, regional or even metropolitan scale, smart or intelligent grids are proposed, capable of two-way information exchange between renewable energy producers and consumers – many of which may combine the same function in the same person, group or company. Here, a much wider range of devices, systems and use modes is coordinated, synchronized, measured and balanced. And within this system, precincts or individual buildings may function autonomously as islands, or as positive contributors to the overall energy supply, exporting into the grid while maintaining their own efficiency management and generation capability.

Figure 1.9 A local power shortage darkens part of São Paolo: A fossil-fuel city on its knees – smart grids may help in this situation but long-term relief can only come in combination with distributed, autonomous renewables, and precinct-based, renewable cogeneration islands. Rolling brown-outs and unscheduled blackouts are growing occurrences in a number of cities and will increase with climate change and fossil fuel supply problems

Source: Rafael Rigues

Cities and their renewable future

Each city is different, and so is its potential to rely on renewable power. There is every indication that the fully renewable re-engineering of existing cities is technically possible. While the empirical evidence is partial and often indirect, it is clear that the opportunities to introduce renewable power into the established urban infrastructure are virtually limitless. The electrification of individual transport alone, only very recently re-emerging as a possible mass reality, offers opportunities of vast proportions.

In order to seize these opportunities, three main elements are helpful. First is a reform or revolution in the institutional arrangements that govern urban decision-making and infrastructure policies and programmes. Second is the empowering move of city agendas from introverted localized or bilateral city–state agendas to multilateral, regional, national and even international strategies in securing and investing in renewable power assets. Investment in distributed, local energy is still key – it affords 70–80 per cent efficiency even when using fossil sources – in contrast to the 30 per cent of traditional, centralized thermal plants (Casten, 2003). And third is the nurturing and deployment of bold civic leadership to act in partnership with community-based organizations, industry and business. Once these elements are put in place and acted upon, nothing can stand in the way of rapid, positive change towards a 100 per cent renewable world.

The world is bound to become renewable

Pre-industrial agrarian communities were founded on fossil fuel free energy sources: manpower, animal strength and the energy in food and feed, plants and trees as biofuel, water, wind – all based on solar radiation. This is still true today; sunlight continues to be the most central and elementary source of life on earth. Even when looking only at the applied energy conversion potential: this outstrips today's conventional commercial energy supply by a factor of 10,000. Fossil fuel, while triggering the great explosion of mechanization known as the Industrial Revolution, may not have been the most essential ingredient in advancing human civilization, compared to the philosophical and scientific breakthroughs leading to it. Renewable systems could have supported equally valuable if different forms of progress: a solar industrial revolution may not have led to intercontinental missiles, mass aviation, industrial fertilizers, DDT, mass production and throwaway society, but many features of progress and change could easily be imagined based on renewable power alone.

While there are still shrinking pockets of more and less sustainable – or attractive – forms of energy independence in remote areas of the world, these have tended to be seen as the impoverished pockets left behind by development and modern civilization. Today, these pockets of relative underdevelopment emerge as frontiers of hope, for leapfrogging over the past stages of what has rapidly become old technology – fossil and nuclear power. This ideal of

Figure 1.10 Boys and solar module, Caoduo school, Rongbo, Yu: Bringing affordable, high-quality solar lighting to rural China as part of the Renewable Energy Development Project (REDP), China

Source: Ashden Awards/Martin Wright

pursuing a 'renewable leapfrog' in developing countries and their cities would be aided if national frameworks and international agreements were reached to ban the promotion and funding of fossil or nuclear fuel-based systems, prioritizing renewable energy-based urban and regional development programmes. This principle should be combined with moves to transcend the structural adjustment programmes of the 1980s, to rebuild local and regional resilience and self-reliance in food production while enabling education, technological and scientific advances and nurture indigenous traditions back to life.

The local case for aiming at the source of emissions

Cities are powerful agents in progress. A lion's share of global emissions can be ascribed to their geographical and administrative territory, and a growing number of programmes focus on cities, from the International Council on Local Environmental Initiative (ICLEI) and its Cities for Climate Protection (CCP) campaign to the Clinton Climate Initiative, or the World Future Council's Cities and Climate Change efforts. Many cities and towns have embraced some form of emissions accounting method. In some countries this

Figure 1.11 One of Bangladeshi NGO Shidhulai's boat libraries visits a remote village in Raishahi, Bangladesh. The promise of technology leapfrogging with solar-powered boats bringing education and sustainable energy to remote areas. Electricity on boats is generated by solar photovoltaic technology

Source: Ashden Awards/Martin Wright

can approach the majority of towns: Australia is one such example where more than half of ICLEI's CCP community was founded during the heyday of the campaign. This has been explained by both a lack of action at the national level, and – as a seeming paradox – central government funding of such programmes. Some cynical voices attribute the extraordinary representation of Australian member cities in CCP to the fact that these local programmes were unlikely to fundamentally change the energy regime's big picture.

The local sustainability movement's emphasis on 'initiative', 'agenda' and 'protection', rather than 'action' or actual 'change' suggests good intentions and agreeable commitments to 'ensuring that future generations are not deprived of choices', or similar aspirations of little meaning – but no real, tangible change. Others call this assessment unfair, pointing to local government's presumed limited range of choices. But the reality of examples suggests otherwise. Quite the contrary, cities and towns are not only quite capable of responding boldly and in a multitude of ways but they actually do so, as summarized below.

There is a set of good reasons for local agendas failing to achieve much progress while focused on emissions policy alone. For one, cities and communities have few means of carrying out long-range planning. It is better to put

Figure 1.12 The clean and worldwide switch to renewable power: Competing with paradoxical, expensive, distant or inefficient carbon reduction schemes such as 'clean coal'

Source: Randy Montoya/Sandia Corporation

in place the best possible and most effective strategies known at any given time to get to zero fossil fuel content in energy use (in transport, household, business and industrial production) and in consumption (embodied energy in the acquisition of goods and services, also known to some as 'grey energy').

Second, after almost two decades of frantic carbon-counting efforts there is no single agreed-upon method emerging, nor even an agreed principle. While the Kyoto-based processes provide a framework of national accounts based on locally generated emissions, to many, this is fundamentally flawed. It does not reflect actual lifestyle or consumption behaviour, and hence does not fairly assign responsibility: much of China's emissions are exported to the US and other importers of Chinese manufacturing goods. And third, the very notion of emission counting and its corollary – trading – is a deeply flawed if popular notion, if only because it does not focus on the sources of emissions, such as fossil fuel combustion, but on its indirect effects, a proximate force. Cities operate below the level of national emission accounts and are not usually equipped to engage in pollution counting programmes in ways that carry much meaning locally or nationally. Often only corporate – city-asset related – emissions have been counted, a paltry fraction of total carbon pollution being emitted within the broad realm of cities and their consumption.

Communities are able to act: Five strategies

Municipal utilities

Direct municipal control over generation proves to be the best way to work towards a renewable portfolio. While in itself it does not guarantee renewable outcomes, having control over local energy production provides much more ability to rapidly respond to the new demands of an increasingly renewable world, while also offering the strength of being able to build up efficient and renewable assets over time, compounding the gains of good practice – instead of being at the whim of generators and distributors without accountability. Two examples may suffice, one in Europe, one in the US. The Bavarian capital of Munich houses Germany's largest municipal enterprise, the Stadtwerke München. This diverse public works and utility operation has consistently invested in innovations as diverse as its venerable district heating networks, geothermal resources, distant offshore wind parks and partnerships in small local bioenergy or waterpower investment funds, or photovoltaic fields. Stadtwerke München disprove the myth that municipal utilities are too costly: in early 2009 municipal energy services in Munich were the second most affordable of the ten largest cities in Germany (SVM, 2009).

One of the ten largest publicly owned utilities in the US, the Sacramento Municipal Utility District (SMUD) was formed in 1923 but was not able to commence operation until 1946, having been fought for decades by commercial operator Pacific Gas & Electric (PG&E), a battle that is still not put to rest. But today, the rise of SMUD as an institution developing and owning energy assets accountable to the community, overcoming internal power struggles, is legendary. One of its celebrated steps was to decommission Sacramento's troubled nuclear reactor, Rancho Seco, from 1997, after 20 years of operation and a dozen years after a serious incident occurred in 1985. This event and other factors gave rise to increased community control over SMUD decisions. Today, almost 40MW of solar power and 500MW of natural gas-fired electricity are generated on the site instead. The SMUD story and others like it also teach how challenging it can be to rebuild enlightened public control, once it has been abandoned.

The privatization waves of the 1980s, when local and state power and water utilities and other public assets were sold to private operators, often international groups and investment schemes, are over. They have given way to the reacquisition of previous offloaded operations, as contemplated by the city of Amsterdam, or reassembled in imaginative ways, such as Danish cooperatives like Copenhagen's alliance to fund, build and manage its offshore wind farm, the Middelgrunden Wind Turbine Cooperative. Others pursue such goals through smaller-area or precinct utilities, or renewable energy service contracting, long used also for efficiency improvements. Sometimes, legal obligations toward large and carbon- or uranium-polluted power providers have to be painstakingly deconstructed, such as when unpopular new coal generators of multinational operators are made to look more efficient by providing district heat as a 'green' by-product of the vast and excessive generation of polluted electricity.

Building virtual utilities

When no power generation assets are owned by the community, energy-related consumption behaviour, efficiency and distributed renewable generation can be guided by public policy innovation, introducing the virtual utility. States and large, powerful metropolitan governments, but also smaller communities with a firm sense of purpose and a modicum of business acumen and corporate discipline, can act to as virtual utilities. Delaware's Sustainable Energy Utility (SEU), has operated since 2008 and has possibly the most comprehensive strategy in the US – chartered to lower all energy consumption by 30 per cent, in transport, businesses and homes by 2015. It also involves an equally comprehensive distributed renewable energy programme – to install 300MW in new renewable capacity by 2019, hoping to reduce carbon emissions by 30 per cent over 2000 figures, in 2020. It is funded by a monthly surcharge of $0.36 on electric bills and a $30 million unsecured bond issue paying dividends on efficiency savings and renewable energy income (Chang, 2008; SEU, 2009).

Building virtual utilities by developing renewable assets

But it is also possible to acquire new assets, a new virtual utility of sorts by directly purchasing generation capability. Munich, the Bavarian capital, had never given up its municipal utility, the Stadtwerke München. München decided to plan, develop and operate solar farms, with renewable company Gehrlicher Solar AG. The aim is to generate in two initial projects 30MW in capacity, part of a strategy to supply all Munich households with renewable electricity (Witt, 2008).

Virtual renewable energy utilities can be created by virtue of public–private partnerships, investing in renewable assets in wind, geothermal or sun energy, acquiring know-how and beginning the process of converting its internal and external power infrastructure into non-fossil, non-nuclear assets, saving billions of dollars in years and decades to come. As an example, off its east coast, the UK not only plays catch-up with its less windy but far more wind-powered neighbour to the south-west, Germany, but also lays the groundwork for a 100 per cent renewable London. Both the London Array, and one of the world's largest wind farms, the 504MW capacity, 140-turbine Greater Gabbard project, are due to commence operation in 2011. They are being developed by Scottish & Southern Energy Plc, with Munich's Siemens AG and Texas-based Fluor as the main contractors, for a total fee of $3 billion shared between both companies (Bergin, 2008).

Setting 100% as a near-term target

The case of Güssing in Austria (*see also* Chapter 6) has been a micro-model of commitment for more than two decades. The town laboured under a massive electricity bill and in the late 1980s began to survey its local and sub-regional renewable energy sources, notably its bioenergy potential in its farming community. It rather quickly succeeded in reaching a 90 per cent emissions reduction level across the community, and promptly attracted a sizeable number of investors and industries in renewable energy and clean technology. Today, the town produces more energy than it consumes – all of it from renewable sources, not unusual for a determined community (for example, the Bavarian town of Wilpoldsried produced 285 per cent of its energy in exports in 2009). This story is repeated across most towns and regions on their path to autonomy – they all set no less than 100 per cent as their target.

Assembling and mobilizing energy city regions

A sound strategy is to form energy compacts across regions and metropolitan areas. They allow energy islands to be linked, resources to be shared, and permit the pooling of resources to invest in regional, distant or even offshore renewable energy projects far beyond the urban boundaries. The Cape Light renewable energy compact to jointly develop and supply certain coastal communities on Cape Cod and Martha's Vineyard, Massachusetts, and its Land-based Wind Collaborative is one of the earlier examples of collaborative, multi-municipal renewable energy. While struggling to expand cape wind resources, Cape Light also provides useful advice on wind development approaches to individual towns (CLC, 2009). Massachusetts is also home to one of the world's largest wind power projects, the 130-generator Nantucket Sound initiative, long hampered by opposition, but finally overcoming it in part in the wake of President Obama's election (Cape Wind, 2009).

The concept and promise of Better Place has prompted Mayor Gary Newsom to announce an alliance with Oakland and San Jose (Metz, 2008). The San Francisco Bay Area is a suitable setting for the introduction of Better Place as a framework for gradually replacing the combustion engine with an electric motor-based car fleet, following the Israeli initiative, Danish plans and an Australia initiative. Better Place is described in Chapter 11.

Community-internal systems supporting a 100% path

Renewable networks: Heat, cooling, power

Rotterdam, Copenhagen and Linz are among the cities best know for highly developed district heating systems. Meanwhile, Linz shows that a reliance on industrial process waste heat is powerful – more than 80 per cent of apartments are connected to a city-wide district heating system fed by the waste heat of industrial areas. But this is also a trap because it discourages the introduction of cleaner energy sources, such as photovoltaic power, and it cements a reliance on the very pollution levels that make waste heat possible. And herein lies Linz's challenge. The waste heat sources can be reduced or eliminated through refurbishment of the offending industries and the building of a renewable base supply infrastructure.

Figure 1.13 Daxu stoves in China: Around 25,000 Daxu stoves have been sold since production started in 2006. The stoves are designed to burn crop waste

Source: Ashden Awards/Martin Wright

Communities can decide to supply biomass-fed or solar-thermal-based district heating systems. A renewably powered electric car fleet can be introduced, using tax incentives and private–public partnership arrangements to facilitate infrastructure and institutional frameworks. And these elements of a renewable world can also best be rolled out in an island fashion. In Germany, for example, cities and communities are encouraged to form linkable heating networks – islands to be connected over time. The state of Baden-Württemberg introduced the country's first renewable heating law on 1 January 2008. One can see the enormous potential: even in Germany, where district heating systems have been relatively common for a long time, renewable sources are still widely underutilized as only some 30 per cent of the wood-based biomass potential, and less than 1 per cent each of geothermal and solar-thermal potential are estimated to be utilized (Staißl et al, 2005).

And to reiterate here once more: distributed co-generation, or combined heat and power, or tri-generation, combining power, heat and cooling, or quad-generation, adding local sewage, mining, water treatment and recycling systems, are powerful ways of increasing system efficiency, introducing renewable energy sources at a large scale and lowering reliance on long-distance transmission networks, which constitute up to half of total hardware construction costs of conventional systems.

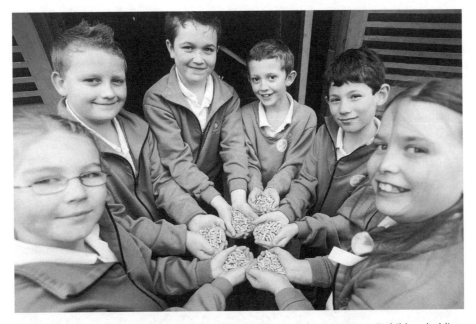

Figure 1.14 Wood fuel heating for schools in Nottinghamshire, UK: Year 5 children holding wood pellets at Mornington Primary School where a new boiler has recently been installed

Source: Ashden Awards/Andrew Aitchison

Distributed generation

In the UK, the city of Woking has been called the local capital of distributed generation, described in Chapter 6. It reached 99.83 per cent autonomy in local electricity supply across its own corporate assets, and maintains a larger grid connection for back-up purposes. The sources, not all renewable, rank from solar to wind, hydrogen, biomass to natural gas. Some 60 generators, from tri-generation plants to solar arrays, heat, cool and power public buildings, affordable housing and city centre businesses. Woking ranks very high on the list of communities that achieved high levels of distributed or micro-generation success. It is funded from the tremendous efficiency savings made over conventional supply, and even attracted Danish pension fund investment. This partial system meant that overall emissions were significantly reduced.

The built environment as energy construct

The hardware of civilization consists of more than clusters of inert containers, at best harbouring an extraordinary potential for energy efficiency savings. Cities and their regions represent huge energy generation, management and distribution opportunities. The pedestrian traffic on pavements, the sunshine and car traffic impacting on streets and roads, the flow of solid waste and liquid refuse, the water stream on surface and in the ground, the flow of air, the solar radiation striking roofs, facades and open spaces, the often enormous geothermal resources that lie both deep and shallow below the surface, when taken together, and combined with ubiquitous point-of-demand and renewably powered heating, cooling and power generation machinery, all are capable of producing at least as much energy as is required to drive and maintain the host precinct.

Figure 1.15 After 10,000 years of evolution, the humble roof has found a new and noble calling, besides keeping out the rain: energy harvesting. Rooftop solar panels in Neckarsulm-Amorbach, Germany

Source: Joachim Köhler

Theoretically even the largest cities could provide their operational requirements, if one disregarded embodied energy in the fabric and flow of consumable goods and services, and included regional energy resources – if mobilization costs could be disregarded. But despite the odds of pushing a fundamentally different approach to energy supply, a growing number of leaders and their communities begin to wonder about that potential, from Spain to China. Teams worldwide have begun to map and calculate the capacity of neighbourhoods, settlements, cities and even regions to convert renewable sources into working energy – adding to the capacity of the business, residential and industrial worlds to innovate, co-generate and distribute energy flows.

The German architect and planner Dagmar Everding has developed a typology of standardized urban patterns prevalent in German cities, calculating their respective potential for converting sunshine into electric and thermal energy (Everding, 2007). Her pattern book has been developed in part with the Dutch-based consulting firm Ecofys, and is used in several cities as a basis for developing renewable energy capacity assessments and local energy plans. Berlin's solar framework plan (Solarer Rahmenplan), too, is an example for her school of precinct-type based solar potential assessments.

Using a more interactive approach geared to individual users, the city of Osnabrück has been the focus of a software pilot project, using satellite data to map the city's solar potential (http://www.osnabrueck.de). Similar efforts are now underway at national government level, experimenting with a set of studies at local level, to better understand the potential of such satellite and aerial data, geographic information systems, and the characteristic patterns of urban settlements, precinct prototypes and open space qualities to assess and map the total generation potential across a wider range of renewable energy technologies, including biomass, geothermal and hydropower. This book features the work of one of the most important protagonist teams in this field in Chapter 14.

National and international infrastructures

Germany has long moved to upgrade its high-voltage grid to take account of massively increased wind power, largely thanks to a super-effective feed-in legislation. This involves not only greater capacity but also the software and systems control and guidance provisions to balance and dispatch a generally predictable but in detail inevitably stochastic source. Early directions and planning are evident in the investigations by Germany's Energy Agency, of the mid-2000s, the so-called dena grid study (dena, 2005).

Supergrids

Several large grid initiatives have emerged since the mid-2000s to shore up international renewable power supply using high-voltage DC over large distances. Most have claimed the prefix 'super', from Airtricity's hydrogen-transmitting and supercooled grid proposal for the North Sea to the plans by the Obama administration for the US. Supergrid concepts are not new, and have in this millennium been aired in venues such as the 2002 National Energy Supergrid Workshop sponsored by the University of Illinois at Urbana-Champagne (UIUC, 2002).

Like smart grids, supergrids are not necessarily conduits for 100 per cent renewable power. Sometimes the latter are proposed as a means to make nuclear power attractive again. The term supergrid has been heard in calls for greater conventional energy security, say, to mitigate Russia's threat to strangle Europe with various power chokes, by patching Baltic, North Sea and Mediterranean grids together, or by implementing alternative natural gas supply rings and access sources. In Europe and across the US, however, supergrids could well help shift great quantities of renewable electricity derived

Figure 1.16 Berlin's solar framework plan

Source: Berlin Digital Environmental Atlas, Urban and Environmental Information System (UEIS), Berlin Department for Urban Development

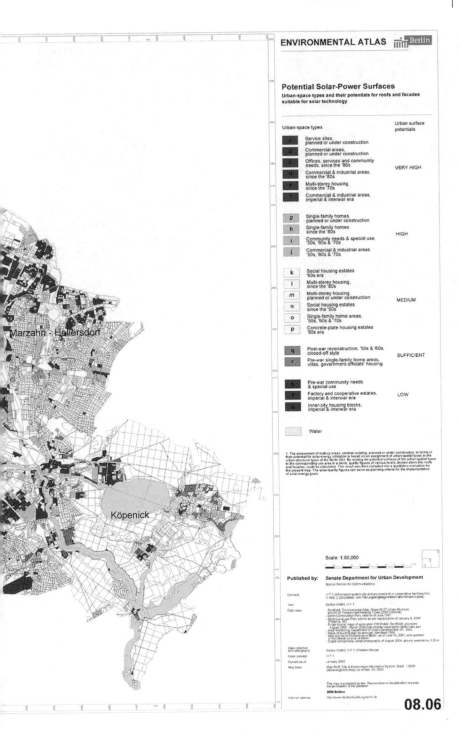

ENVIRONMENTAL ATLAS ⬚ Berlin

Potential Solar-Power Surfaces
Urban-space types and their potentials for roofs and facades
suitable for solar technology

Urban-space types		Urban surface potentials
a	Service sites, planned or under construction	
b	Commercial areas, planned or under construction	
c	Offices, services and community needs, since the '80s	
d	Commercial & industrial areas, since the '80s	VERY HIGH
e	Multi-storey housing since the '70s	
f	Commercial & industrial areas, imperial & interwar era	
g	Single-family homes, planned or under construction	
h	Single-family homes since the '80s	
i	Community needs & special use, '50s, '60s & '70s	HIGH
j	Commercial & industrial areas '50s, '60s & '70s	
k	Social housing estates '60s era	
l	Multi-storey housing, since the '80s	
m	Multi-storey housing planned or under construction	MEDIUM
n	Social housing estates since the '50s	
o	Single-family home areas, '50s, '60s & '70s	
p	Concrete-plate housing estates '80s era	
q	Post-war reconstruction, '50s & '60s, closed-off style	SUFFICIENT
r	Pre-war single-family home areas, villas, government-officials' housing	
s	Pre-war community needs & special use	
t	Factory and cooperative estates, imperial & interwar era	LOW
u	Inner-city housing blocks, imperial & interwar era	
	Water	

1. The assessment of built-up areas, whether existing, planned or under construction, in terms of their potential for solar-energy utilization is based on an assignment of urban-spatial types to the urban-structural types of the Berlin ISU. By relating the potential surfaces of the urban-spatial types to the corresponding use area in a block, quality figures of various levels, broken down into roofs and facades, could be calculated. This result was then compiled into a qualitative evaluation for the present map. The solar-quality figures can serve as planning criteria for the implementation of solar-energy goals.

Scale: 1:50,000

Published by: **Senate Department for Urban Development**
Special Section for Communications

Concept: I&F 1 (Information system city and environment) in cooperative SenGesUmV
II A/IC 2 (Grundsatz- und Planungsangelegenheiten des Klimaschutzes)

Text: Ecofys GmbH, I&F 1

Data base: - SenStadt, Environmental Atlas, Maps 06.07 Urban Structure
and 06.10 Predominant Heating Types (2005 Edition)
- Berlin Construction Plan, valid as of June 1961
- Berlin Land-use Plan, edition as per republication of January 6, 2004
(Präent p. 95)
- Single-layout maps of application FIS Broker, SenStadt, accessed
August 2005 - March 2006 (http://fbinfer.stadt-berlin.de/fb/index.jsp)
- Area Monitoring, Department of Urban Development (A?, 2005)
- Maps of building age, by borough, SenStadt 1999
- Map and list of monuments in Berlin, as of June 14, 2001, and updated
in the Official Journal of Berlin
- Digital orthophotos, aerial photography of August 2004, ground resolution 0.25 m

Data collection and cartography: Ecofys GmbH, I&F 1, Christian Bloßer

Color concept: I&F 1

Current as of: January 2008

Map base: Map ISU5, City & Environment Information System, Scale 1:5000
(block-segment map), as of Dec. 30, 2005

This map is protected by law. Reproduction or republication requires
the permission of the publisher

2006 Edition

Internet address: http://www.stadtentwicklung.berlin.de

08.06

Marzahn - Hellersdorf

Köpenick

Figure 1.17 Chicago streets

Source: David Niblack

from wind and solar power from Nevada and Arizona to other states, or from Scotland's enormous tidal, wave and wind resources across Europe, as part of the European North Sea Offshore Grid initiative. An agreement between Norway and The Netherlands follows a much earlier agreement with Denmark to use Norwegian water resources to balance the variability of the wind (see www.terrawatts.com; http://business.timesonline.co.uk/tol/business/industry_sectors/natural_resources/article5142622.ece). Going further from 'super' to 'super smart' in the battle of the self-bestowed grid accolades, a coalition of large power utilities and environmental groups including WWF and Vattenfall Europe Transmission announced on 3. July 2009 the Renewables Grid Initiative (RGI) to promote the integration of distributed renewables sources. The RGI is founded on the ideas of Antonella Battaglini, Process Leader of the SuperSmart Grid Project in the European Climate Forum (see http://www.renewables-grid.eu; www.european-climate-forum; www.pik-potsdam.de).

Also across Europe, the earlier and controversial Trans-Mediterranean Interconnection for Concentrating Solar Power (TRANS-CSP) project was aimed at interconnecting the electric grids of Europe and North Africa, with the aim of importing solar energy from solar thermal (CSP) sources in North Africa, at a rate of some 15 per cent of Europe's electricity demand by 2050. An early planning study was funded by Germany's Federal Ministry for Environment, Nature Conservation and Nuclear Safety (BMU) and completed by an eight-member consortium comprised of German, Egyptian and Algerian firms and agencies, led by the German Aerospace Center (DLR). The study found, based on modelling for 30 European countries from 2000 to 2050, that a full renewable power supply was possible, indeed, there is a capacity of some 145 per cent of conventional electricity supply supplied through a grid of pan-European biomass, hydropower, geothermal and solar energy assets, linked through a new DC grid supplementing an upgraded alternating current (AC) network (Trieb et al, 2006). Critics point out the many environmental, political and efficiency difficulties with schemes so founded on the big infrastructure dreams of the late 19th century, while diverting much needed funds for locally integrated and controlled sources. Undeterred, the DESERTEC Foundation emerged from TRANS-CSP, with its well-heeled dream to supplement European renewable renewable energy sources with DC-wired desert sun courtesy of North Africa (http://www.desertec.org; http://eurosolar.de/de/index.php?option=com_content&task=view&id=1161&Itemid=324).

First step to a supergrid: Australia's renewable energy backbone

Supergrid elements can be built gradually. Australia's private sector proposed Inland Electricity Transmission Connection is an example of a long-mooted simple DC feeder, linking across 1200 kilometres (km) of Australian inland territory, considered largely desert, but rich in minerals and now suddenly appreciated at a commercial scale – hot dry rock geothermal resources, wind and sun exposure. Poetically dubbing these vast potential wind, solar thermal and geothermal resources as 'the National Energy Market's (NEM) Inland Energy Farm', entrepreneurs call for the government to unlock it by introducing a 1200km 350 kilovolt (kV) DC line with 300MW capacity, linking the northeast of South Australia with the south-western corner of Queensland, and their respective AC grids (Evans and Peck, 2008).

Smart grids

Grid upgrades have long gripped the imagination of the electricity distribution community, software world, control industry and political leaders alike. Power networks have long been neglected in infrastructure thinking, stuck very much in early Tesla time – the pioneering days of wired electricity. In the developing world, too, 'electrification' has meant the stringing of high-tension wires over long distances, hooking communities to the gargantuan, wasteful guzzlers of cooling water that were thermal reactors fuelled with uranium, coal, gas or even oil. Today, a race is underway to improve the capability of networks: to respond to failures more quickly, maintain interaction between customers and suppliers to manage failures, allow home reading of live usage and facilitate the two-way flow of electricity.

Figure 1.18 Old-fashioned power and telephone lines sagging after heavy ice storm. New integrated services power and information networks (the 'e-web'; see Droege, 2006) and increased embedded, distributed, stand-alone RE systems mitigate climate change and enhance resilience

Source: noaa.gov

These relatively mild enhancements came much later than the revolutions in the telecommunications world, and only after the rise of terrorist threats and severe power outages struck California and the northeast of the US in the early to mid-2000s. Measures included software innovations, control hardware and management reforms. Initially also conceived to learn about smaller power failures and enhance customer understanding of efficient behaviour, they have now given way to more serious, if increasingly Orwellian, thoughts about managing a wide array of diverse inputs from small solar panels to massive wind parks, municipal-scale solar-thermal fields or wave power stations – and connect to a myriad of storage devices, from traditional pumped dams to stationary chemical, mechanical or electrical batteries and devices, to mobile storage in electric vehicles and plug-in hybrids. A truly smart grid would also facilitate 100 per cent renewable power and include virtual network features, local area management and interactive trades across a number of communities – the *e-web* (Droege, 2006).

Demand-driven improvements, long resisted by the power carriers and fund-starved managers of public assets, inevitably took on features of technology convergence familiar from the heyday of the early information revolution: the merger and partial crossover of voice, data and image media. It

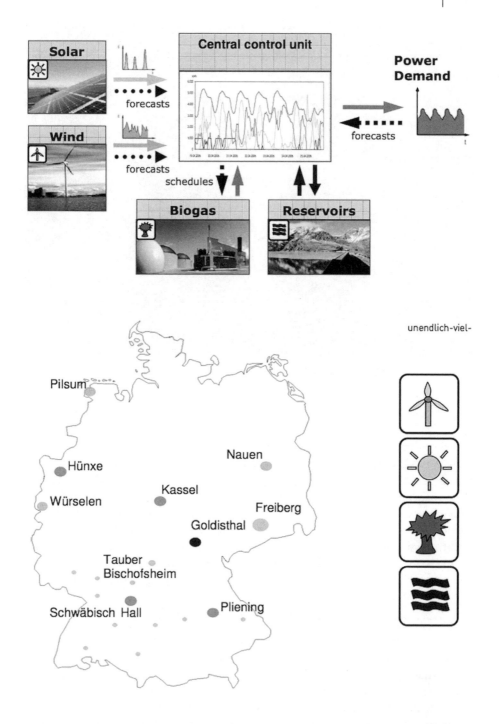

Figure 1.19 A 1:10,000 scale working model of an autonomous renewable electricity supply system for Germany

Source: Agentur für Erneuerbare Energien e.V. www.unendlich-viel-energie.de / kombikraftwerk.de

Figure 1.20 Local distribution heritage: Street wire tangle

Source: Darrell Rogers

is little surprise that the most imaginative players include software moguls like Shai Agassi, initiator of Better Place, the electric car package; IBM and its Smart Grid initiative; and Google who joined the Californian Demand Response and Smart Grid Alliance in November 2008, releasing *Google PowerMeter* in February 2009, a software application aimed at the home-based smart grid user (see http://news.cnet.com/8301-11128_3-10160234-54.html; www.smartgrids.eu/; http://smartgridnews.com; http://money.cnn.com/news/newsfeeds/articles/marketwire/0477001.htm; www.ge.ecomagination.com/smartgrid; www.oe.energy.gov/smartgrid.htm).

Not 'super' but smart: The arrival of the virtual power plant

The year after the old TRANS-CSP study was completed, three German companies, Enercon, Schmack Biogas and Solarworld, under the auspices of the Berlin-based Agency for Renewable Energies, released the results of an experiment they conducted to demonstrate the ability of renewable energy to supply all of the electricity requirements of Germany. They linked 36 renewable energy sources and management points across the country: a pumped-water storage facility for general balancing, 4 biogas block heating plants, 11 wind parks and 20 solar power plants. Daily weather prognoses were used to gauge general demand and supply balances and real-time data to fine-tune supply with up-to-the minute accuracy. Because biogas is also a storage medium, the block heating units served as peak-shaving and not base load suppliers, and hence the main balancing tool to manage fine changes in the supply-demand profile. The experiment was scaled to represent 1:10,000 of Germany's total electricity demand, supplying the equivalent of nearly 18,000 households.

Repowering America: A call for a renewable national power grid

The vision of a fully fossil and nuclear free world has also inspired some progressive leaders on these issues: for example, Al Gore presented the US Congress on 17 July 2008 with 'A Generational Challenge to Repower America', to move to 100 per cent renewable electricity within ten years. The four pillars of this proposed programme are no surprise: greater energy efficiency (difficult to achieve in a decade, except for new assets); renewable energy generation such as solar, wind and geothermal power; a national power grid integration programme involving both continental and local smart grid networks; and, as an integral vision to an advanced grid, the shift away from dumb internal combustion engines to reformed national assembly lines producing intelligent electric vehicles. Their collective battery power offers an enormous balancing capacity for renewable energy resources (TA, 2008).

Postscripts

These national efforts, especially those driven by incumbent industry agendas, only reinforce the need for urban and community-based action. National and international agendas remain uncertain. The popular hunger for a renewable world underpinned, at least in part, Barack Obama's 2009 election win: he had embraced a renewable energy supportive agenda to be part of his election promises. But it is wise to remember that he joined other US leaders before him, including President Jimmy Carter, who already in 1979 had solar hot water heaters installed in the White House, and announced his 'solar strategy'. Under Ronald Reagan the panels were removed, and after an odyssey that included their symbolic hosting in Unity College near Bangor, Maine, they have now become a display item in the Jimmy Carter Presidential Library (Clark, 2007).

In installing the solar water heating system, Jimmy Carter identified a generation ago a civilization at a crossroads, and we are still lingering indecisively at this very crossroads today. Carter declared:

> a generation from now, this solar heater can either be a curiosity, a museum piece, an example of a road not taken, or it can be a small part of one of the greatest and most exciting adventures ever undertaken by the American people; harnessing the power of the Sun to enrich our lives as we move away from our crippling dependence on foreign oil.

A year later, US domestic oil production capacity peaked, as had been predicted by Shell geophysicist Marion King Hubbert a generation earlier (Hubbert, 1956) yet the road was abandoned, and again under Bill Clinton and Al Gore's time in office. 1980 is also the year when the Federal Emergency Management Agency issued its futile call for a renewable America (FEMA, 1980).

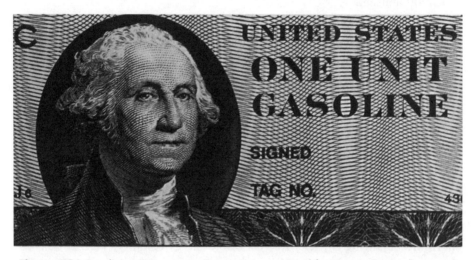

Figure 1.21 Gasoline ration coupon: Coupons were printed for emergency use (but never issued) during the energy crisis in 1979

Source: United States Department of Energy

While today a different awareness has arisen, with the spectre of catastrophic global warming etched into growing parts of the public (sub)conscience, adding to the managed energy security dangers that marked the early and mid-1970s, it is wise to not entirely rely on national or international processes alone.

A prospect: What future for air traffic?

Aviation can be renewable, too, but whether mass air travel can or should possibly be sustained without a landing at saner levels is another question, certainly in the context of its evolution in the 1990s and 2000s. Solar, wind, geothermal or oceanic power hydrogen production plants could one day be ready to power airplanes, while biofuel offers extremely marginal prospects in the short-term only – such as attempts to combine landfill waste, sorghum, wood chips and other sources into aviation biofuel (McDermott, 2008). Indeed, plant-based fuel production is far more inefficient than solar thermal or electric conversion, and often competes with food production. Excessive air travel, and aviation more broadly, must inexorably be corrected downward in a healthy way. So-called Airotropolis (airport city) schemes (Kasarda, undated), urban development programmes banking on airport expansions and associated business development, belong to the energy-blind days of yesteryear. Instead, regional development and re-localization are the call of the day, with a far more selective application of airborne modes of conveyance.

... and a caveat: Does renewable always mean sustainable?

Figure 1.22 Aerial view of Levittown, Pennsylvania, circa 1959

Source: National Archives and Records Administration

Solar, wind and many other renewable sources are ubiquitous in their generation potential. Others, such as large wave power fields, or geothermal sources can provide very substantial amounts of energy, but only in specific locations. Both often benefit from being used locally, saving the installation of expensive and sometimes extensive grid connections for new industrial or residential developments. But this very benefit removes one more important constraint on urban sprawl. Indeed, the lifting of constraints inherent in urban infrastructure innovations has had profound and not always positive implications for the form of cities and their impact on natural or rural areas. Cities have spread and become global, virtual network features, from horse-and-buggy to the streetcar, electrified suburbs and the motorcar, telecommunications and mass air travel (Droege, 2006). Urban sprawl is considered a major cost of the fossil fuel era and its car dependent, heavy infrastructure-based ideas. While most of these innovations were driven by petroleum-enabled technologies, there is nothing inherently healing about renewable energy when it comes to urban sprawl. The spread of solar suburbs and off-grid vacation homes in the US's south-east, such as Arup Engineer's solar expansion of 300,000 inhabitants in Phoenix, Arizona (Boyd, 2007), or the development of cottages in the 30,000 islands of Georgian Bay, Canada, raise important challenges. And so do large-scale, renewably themed development initiatives – many endowed with global brand name designer and engineering credentials – unless these became embedded in larger, authentic renewable city regeneration efforts, and in many cases, become replaced by these.

The application of renewable energy in itself does not always alter the logic of development. Its power to both disconnect urban development from existing cities and to sometimes be used as a promotional argument to accept certain other environmental or social costs only amplifies the need to carefully push and plan the integration with existing centres, with the preservation and renewable regeneration of existing urban areas and with the un-building or un-development of potentially healthy natural areas, not seeing 'renewable' as always synonymous with 'sustainable'.

While a 100 per cent renewable energy base alone is not sufficient for sustainable practice, it is its *conditio sine qua non* – sustainability's necessary condition. May this book be a contribution towards achieving true sustainability founded on a fossil fuel and nuclear free reality.

Figure 1.23 Vision of the Western Australian capital city, Perth, supported by photovoltaic or solar-thermal fields anchoring a diverse city region developed around urban agriculture, forest and wetland resources

Source: Weller (2009)

Conclusions

In a summary conclusion, the successes and potentials represented in this book demonstrate that a move to a 100 per cent renewable civilization is not only required but indeed possible.

Signs of a great shift are visible everywhere, particularly where people live and work (and where a lion's share of global fossil energy is consumed) in cities. These include:

- resuscitated efficiency programmes such as those reinforced by the Clinton Climate Initiative, and many national efforts, from the Swiss Minergie framework, to German efficiency legislation, or the reborn state-based and locally focused US megawatt drives;
- demand reduction efforts in transport such as the compact-city and public transport movements, and less frequently, in lifestyle changes such as urban agriculture and re-localization initiatives (including those that take place on the White House lawn);
- urban-integrated renewable energy systems, such as those assisted by feed-in tariffs: building integrated photovoltaics, distributed renewable cogeneration plants, small wind and hydropower, or geothermal energy;
- regional renewable supply in solar, wind, hydro, geothermal and biomass use;
- system-penetrating sectoral transformations, such as in moves towards entirely renewable transport means, such as a solar powered and shared electric vehicle fleets;
- the rising realization that the MDGs can by definition only be attained in a developing and industrialized world that is founded on a fully renewable energy base.

References

ALTER – Le Groupe de Bellevue 1978 (1996) 'A study of a long-term energy future for France based on 100% renewable energies', in The *Yearbook of Renewable Energies 1997/96*, Ponte Press, London

Archer, C. L. and Jacobson, M. Z. (2005) 'Evaluation of global wind power', *Journal of Geophysical Research*, vol 110, D12110, www.agu.org/pubs/crossref/2005/2004JD005462.shtml

Bergin, T. (2008) 'World's largest offshore wind farm in the works', *Reuters Environment News*, 14 May, www.reuters.com/article/environmentNews/idUSL1483748320080514

BMU (Federal Ministry for the Environment, Nature Conservation and Nuclear Safety) (2008) 'Renewable energy sources in figures: National and international development', BMU, Berlin

Boyd, O. (2007) 'Arup to design world's first "solar city" in Arizona', *Building*, 27 November, www.building.co.uk/story.asp?sectioncode=284&storycode=3100805&c=2

Cape Wind (2009) 'America's first offshore wind farm on Nantucket Sound', www.capewind.org/

Casten, T. R. (2003) 'Thinking outside the box: Economic growth and the central generation paradigm', USAEE/IAEE, http://www.masstech.org/renewableenergy/public_policy/DG/ resources/2003-08-04-Tom-Casten-USAEE-IAEE.pdf

Chang, S. A. (2008) 'The rise of the energy efficiency utility', www.spectrum.ieee.org/may08/6216

Clark, T. (2007) 'White House solar panel goes on display at Carter Library. 1979 effort to encourage alternative energy sources became "Road not taken"', Jimmy Carter Library & Museum News Release, 27 March, www.jimmycarterlibrary.org/newsreleases/2007/07-18.pdf

CLC (Cape Light Compact) (2008) 'Powering Cape Cod and Martha's Vinyard', www.capelightcompact.org/ and www.capelightcompact.org/landbased_wind.html

dena (2005) 'Planning of the grid integration of wind energy in Germany, onshore and offshore up to the year 2020', Deutsche Energie-Agentur GmbH, www.wind-energie.de/fileadmin/dokumente/Themen_A-Z/Netzausbau/stud_summary-dena_grid.pdf

Droege, P. (2006) Renewable City: Comprehensive Guide to an Urban Revolution, Wiley, Chichester

EREC (European Renewable Energy Council) and Greenpeace (2007) Future Investments Study, EREC and Greenpeace, Brussels, http://www.greenpeace.org/raw/content/sweden/rapporter-och-dokument/energirevolution.pdf

Evans&Peck (2008) 'Inland electricity transmission connection. Submission to Infrastructure Australia/Allocation from the Building Australia Fund', www.infrastructureaustralia.gov.au/public_submissions/published/

Everding, D. (ed.) (2007) *Solarer Städtebau*, Kohlhammer

FEMA (Federal Emergency Management Agency) (1980) *Dispersed, Decentralised and Renewable Energy Sources: Alternatives to National Vulnerability and War*, FEMA, Washington, DC

Hansen, J., Sato, M., Kharecha, P., Beerling, D., Berner, R., Masson-Delmotte, V., Pagani, M., Raymo, M., Royer, D. L. and Zachos, J. C. (2008) 'Target atmospheric CO_2: Where should humanity aim?', *Open Atmos. Sci. J.*, vol 2, pp217–231, http://arxiv.org/abs/0804.1126

Homer-Dixon, T. (2009) The Upside of Down: Catastrophe, Creativity, and the Renewal of Civilization, Resource & Conflict Analysis, Inc. and Island Press, www.worldwatch.org/node/6008

Hubbert, M. K. (1956) *Nuclear Energy and the Fossil Fuels*, No 95, Shell Development Company, Houston, www.hubbertpeak.com/hubbert/1956/1956.pdf/

Johnson, P. (2006) *Ecological Footprint Accounting for Western Australia: Technical Paper No. 4*, Environmental Protection Authority, Perth

Jordan, A. (2009) 'Reno: America's biggest little geothermal city', Wall Street Journal, 5 March

Kasarda, J. D. (undated) 'Airotropolis', www.aerotropolis.com/aerotropolis.html

Lehmann, H. (2003) 'Energy-rich Japan', Institute for Sustainable Solutions and Innovation (ISUSI), Aachen

Lenzen, M., Wood, R. and Barney Foran. (2008) *Direct versus embodied energy: the need for urban lifestyle transitions*, in Droege, P. (ed) Urban Energy Transition, Elsevier, London

Lohman, L. (ed.) (2007) 'Carbon trading: A critical conversation on climate change, privatisation and power', www.dhf.uu.se/pdffiler/DD2006_48_carbon_trading/carbon_trading_web_HQ.pdf

McDermott, M. (2008) 'Greener flying?', www.treehugger.com/files/2008/08/swift-enterprises-develops-bio-aviation-fuel.php

Meadows, D. H., Meadows, D. L. and Randers, J. (2004) *Limits to Growth: The 30-Year Update*, Earthscan, London

Metz, C. (2008) 'San Francisco enters Agassi's electric car dream', www.theregister.co.uk/2008/11/21/san_francisco_better_place/

Mills, D. R. and Morgan, R. G. (2008) 'Solar thermal electricity as the primary replacement for coal and oil in US generation and transportation', http://ausra.com/pdfs/ausra_usgridsupply.pdf

Newman, P., Beatley, T. and Boyer, H. (2008a) Resilient Cities -Responding to Peak Oil and Climate Change, Island Press, Washington, DC

NOAA (National Oceanic and Atmospheric Organization) (2009) 'Trends in Atmospheric Carbon Dioxide: Mauna Loa', www.esrl.noaa.gov/gmd/ccgg/trends/

Peter, S. and Lehmann, H. (2004) 'Das deutsche Ausbaupotential erneuerbarer Energien im Stromsektor', Institute for Sustainable Solutions and Innovations (ISUSI), Eurosolar, Aachen

Rahmstorf, S. and Schellnhuber, H. J. (2007) *Der Klimawandel. Diagnose, Prognose, Therapie*, C. H. Beck

Rotman, G. (2009) 'Gilber Metcalf: The case for a carbon tax', *Technology Review*, vol 112, no 1, p28

Scheer, H. (2002) *The Solar Economy – Renewable Energy for a Sustainable Global Future*, Earthscan, London (iriginal in German, 1999)

Schindler, J. and Zittel, W. (2007) 'Crude oil: The supply outlook', Ludwig-Bölkow-Systemtechnik GmbH, Munich

Schreuder, Y. (2009) *The Corporate Greenhouse: Climate Change Policy in a Globalizing World*. Zed Books, pp213–214

SEU (Delaware Sustainable Energy Utility) (2009) 'Sustainability Energy Utility, Oversight Board', http://www.seu-de.org/

Slavin, T. (2007) 'Charging up the councils. Green Futures', www.forumforthefuture.org/greenfutures/articles/ESCO

Staiß, F., Böhnisch, H. and Krewitt, W. (2005) 'Der Wärmemarkt – Analysen und Potenziale erneuerbarer Energie-quellen', paper presented at Jahrestagung Forschungsverbund Sonnenenergie, 22–23 September, Cologne

SVM (2009) 'SWM Preise weiterhin günstig', Stadtwerke München, www.swm.de/de/produkte/swm-preiswert.html

TA (Thai Automaxx) (2008) 'Renault-Nissan to build electric car in Denmark', www.thaiautomaxx.com/2008/07/renault-nissan-to-build-electric-car-in.html

Trieb, F. et al (2006) 'Trans-Mediterranean Interconnection for Concentrating Solar Power', German Aerospace Center (DLR) Institute of Technical Thermodynamics Section Systems Analysis and Technology Assessment. Study commissioned by Federal Ministry for the Environment, Nature Conservation and Nuclear Safety, Germany, www.dlr.de/tt/trans-csp

USGS (United States Geological Survey) (2005) 'Central Region Energy Resources Team: Worldwide web information on United States Energy and World Energy Production and Consumption Statistics', http://energy.cr.usgs.gov/energy/stats_ctry/Stat1.html#WProduction/

UIUC (2002) 'National energy supergrid workshop report', http://energy.ece.illinois.edu/SuperGridReportFinal.pdf

Weller, R. (2009) *Boomtown 2050: Scenarios for a Rapidly Growing City*, University of Western Australia Press, Perth

Witt, A. (2008) '(Stadtwerke) München investieren 1 Milliarde', Solarthemen, 5 July, www.reuters.com/article/environmentNews/idUSL1483748320080514

Zweibel, K., Mason, J. and Fthenakis, V. (2008) 'A solar grand plan', *Scientific American*, www.sciam.com/article.cfm?id=a-solar-grand-plan

Chapter Two

Institutions for a 100% Renewable World

Hermann Scheer

Mandated by governments worldwide, IRENA aims at becoming the main driving force in promoting a rapid transition towards the widespread and sustainable use of renewable energy on a global scale.

Acting as the global voice for renewable energies, IRENA will provide practical advice and support for both industrialized and developing countries, help them improve their regulatory frameworks and build capacity.

The agency will facilitate access to all relevant information including reliable data on the potential of renewable energy, best practices, effective financial mechanisms and state-of-the-art technological expertise.

(Federal Ministry for the Environment,
Nature Conservation and Nuclear Safety, 2008)

The future of power lies with renewable energies. The limits of fossil and nuclear energy are more than obvious. Civilization stands at a critical decision point. The global community can continue down the path to self-annihilation by wasting trillions of precious funds in oil drilling, shale, tar sand and frozen methane production, and pursuing hopeless nuclear fission and fusion research. Or it can end the madness of a bygone era and focus its remaining resources on a strategy of survival and prosperity by building an efficient, equitable and sustainable power infrastructure based on renewable energy.

Recognize the limits in order to overcome them

The first limitation of the conventional power system is physical. The energy demand of a growing world population increases at a faster pace than the gains in energy efficiency and conservation. Mineral resources are limited. Every thinking person understands that oil, gas, coal and uranium reserves are finite, but not everyone yet understands that production capacity is very likely to already be in decline today – while demand continues to soar. This inexorably results in spiralling energy prices, supply shortages in many national economies and social problems for an increasing number of countries and their citizens. Access to energy has become a global political issue. But as long as all eyes are on the old paradigm of power control there is little hope of transcending this dreadful policy and action conundrum, this state of paralysis. The call for 100 per cent renewable energy is essential to help us focus on the far more advanced, essential new energy paradigm.

The direct costs of conventional energies can only rise while those of renewable energies can only fall. Renewable energies are by definition in infinite supply and, with the exception of biomass, their primary source is free. Costs for the production of energy deriving from renewable sources include the required technologies, the hardware and services associated with it, but not for fuels. Only biomass-derived energy creates source costs due to the agricultural, forestry and other inputs required to grow, manage, harvest and process plants. The cost of technology falls due to economies of scale and the predicted rise in the productivity of the deployed technologies, which are still comparatively young. Today's higher costs for renewable energies, where these still apply, are essential for an economically viable future energy supply, available everywhere and for everyone. This promising future is closer than most people think, or would have us believe, particularly those who have ignored or underestimated the potential of renewable energies. Among these culprits are governments, scientists and dominant sections of the conventional energy sector.

The second limitation imposed by the conventional system of energy supply is ecological. Even if vast new oil, gas or coal reserves were to be found, world civilization could ill afford their use. The ecosphere's capacity to mitigate damages has already been breached. The switch to renewable energies has to occur now – long before fossil fuels are depleted. The window of effective action may be as small as ten years, perhaps less. We are in a race against time.

But even if manmade global warming or fossil fuel depletion did not exist, the global energy system would still not be healthy. Their environmental, social and economic costs are enormous. Current energy prices do not reflect these costs – but they are being paid nonetheless. Only renewable energy can liberate society from these shackles.

And yet there are those who regard any sensible response to this existential challenge as an economic burden. This argument is built upon a short-sighted fallacy that has been long unmasked but that continues to cast a heavy shadow on the current energy discussion, in this lingering climate of so-called economic rationalism. The switch to renewable energy promises a number of powerful political, economic, social and ecologic benefits, many of them quantifiable. These are usually overlooked in the laser-like focus at microeconomics, or in the terribly limited and insular cost comparisons of various energy investments. A macroeconomic, comprehensive view leads to a dramatically different understanding.

Yet while macroeconomic benefits are powerfully evident, they cannot deliver microeconomic benefits for every player in the national economy. Well-informed and far-sighted political measures and instruments are mandatory to translate macroeconomic benefits into microeconomic gains and incentives. A good example of this principle is the German renewable energy sources or feed-in tariff law. Since renewables so clearly have macroeconomic benefits for society as a whole, they have been supported by law in Germany, initially in the production of electricity. Guaranteed grid access for renewable electricity, a guaranteed FiT – without cap – dramatically lowers the investment risk for renewable energy producers. This law abolished market barriers stimulating investments effectively.

While it has been obvious for some time that renewable power is essential for a survivable future, most countries are not very well prepared for the inevitable transition. It began to dawn on world governments only recently that renewable policies have to be focused on and promoted. Hence implementation lags massively behind. Many countries encourage the production and use of renewable energy at political and economic levels, but woefully few have drafted and implemented any ambitious policies so far, with the necessary scientific, technological and industrial prerequisites at their disposal. It is no small wonder: the limitless sources of the sun have been marginalized effectively and methodically rendered irrelevant in the global energy discourse throughout the 20th century.

Institutionalizing energy innovation after World War II

In the 1940s and 50ies, energy policy focus began to be trained on nuclear power, in the US starting with the founding of the Atomic Energy Commission (1947) and President Eisenhower's Atoms for Peace programme (1953). The attitude towards nuclear energy then was the opposite of how renewables were treated until recently: potentials were wildly overstated and the risks woefully underestimated. Virtually all industrialized countries of any ambition felt compelled to bias their national energy strategies towards nuclear. To support this trend, two international institutions were established in 1957: EURATOM in Western Europe and the International Atomic Energy Agency (IAEA), with its global focus. The establishment of the latter was welcomed by the UN but not embraced as part of the UN family. In 1956, 82 UN member states negotiated the Treaty that entered into force the following year (Fischer, 1997).

The IAEA is not only charged with preventing the abuse of fissile material, but also carries the mandate to help governments develop nuclear energy programmes, to facilitate technology transfer and build human resource capacities. Yet atomic energy's star, once shining so bright, has long been eclipsing, even if the industry refuses to accept this. The IAEA, half a century old, does well in this self-perpetuating demi-world, with some 2200 employees (http://www.iaea.org/About/staff.html) and an annual budget of more than $250 million.

The quest for an energy agency of the future

Renewable energies represent the very future of global energy supply and yet no adequate agency was created to promote their spread. This glaring imbalance between societal demand and policy support alone provides a powerful motive for setting up an agency chartered with the massive spread of renewables: the International Renewable Energy Agency (IRENA). The call to establish such an agency was raised for the first time 28 years ago in the context of the North–South Commission's Report chaired by former German Chancellor Willy Brandt. The establishment of IRENA was recommended in the final resolution of the first UN conference on renewable energy in Nairobi in 1981, the Conference on New and Renewable Sources of Energy. Nevertheless, these recommendations remained largely unheeded. It was argued that it would suffice to entrust existing UN organizations with the promotion of renewable power.

Yet the need to squarely focus on promoting renewables internationally grew steadily. The 1973 oil crisis showed plainly that the oil age would not last forever. To primarily help monitor and manage security of fossil supplies, the OECD countries established the International Energy Agency (IEA) in 1974, called for by Henry Kissinger a year earlier: 'the answer could only be... a massive effort to provide producers an incentive to increase their supply, to encourage consumers to use existing supplies more rationally and to develop alternative energy sources' (Kissinger, December 1973, cited in Parra, 2004).

Because of its focus on the needs of industrialized, largely oil-consuming countries, the IEA did not evolve into a UN agency either – it was soon regarded as a 'Club of the Rich'. After the European Atomic Energy Community (EURATOM) and the IAEA, a third international organization covering energy matters had thus been established. All three maintain powerful industry and government links – part of a dangerous collusion to exclude renewable energy from mainstream discourse and policy platforms.

Although most industrialized nations announced initial R&D programmes for renewable energy after the oil crisis, the priority of R&D funding lay elsewhere. When oil prices declined in the early 1980s, most countries scaled back their nascent renewable initiatives. This soon triggered unrest. The 1980s and 1990s witnessed a growing and widespread unease about the mounting nuclear and fossil energy dependence, its risks and its costs. The Chernobyl disaster in 1986, the Three Mile Island near-melt down in 1978, a year after the plant's commissioning, and a series of other mishaps combined with the madness of the atomic arms race to compound the strong resistance to nuclear power. The 1990s, with climate change reports growing increasingly alarming, saw a further surge in criticism of the fossil energy conundrum. But these calls reached the mainstream international energy discussion terribly late, so entrenched was the belief that there would not be a realistic alternative to conventional energies.

IRENA rising

To help counter this myth, various scientific studies were conducted to show that a complete energy supply with renewables would be feasible. Examples include a report by the Union of Concerned Scientists in the United States in 1979 (Kendall and Nadis, 1980); a publication of the Club de Bellevue, an initiative of scientists from leading French research institutes (ALTER, 1978); or a Europe-wide study released by the Institute of Applied Systems Analysis in Laxenburg (Austria) in 1982 (www.iiasa.ac.at/Admin/PUB/Documents/WP-82-126.pdf). The technical capacity to transform the global energy system clearly existed, the societal need clearly existed, and yet there were no international policy sources or high-level advocates to help bring about choices and pave the way for a massive shift towards renewables.

In 1990, the European Association for Renewable Energies, Eurosolar, drafted the first comprehensive memorandum on establishing IRENA, publishing it widely. At the invitation of the former energy commissioner of the UN Secretary General, Ahmedou Ould-Abdallah, I presented this memorandum at the UN headquarters in New York. UN Secretary General Perez de Cuellar responded by establishing a task force, UNSEGED (United Nations Solar Energy Group on Environment and Development). UNSEGED, chaired by Professor Thomas Johansson, concluded that the establishment of an International Renewable Energy Agency was necessary. This proposal was aimed at the Rio Conference of 1992 and it was expected that this conference

Figure 2.1 Delegates at the International Parliamentary Forum on Renewable Energies, June 2004, Bonn, agreeing on the final resolution

Source: Aribert Peters

would establish the agency. At the invitation of the US Senate, the Interparliamentary Conference on the Global Environment took place in Washington DC in 1991, chaired by Senator Al Gore. At this conference, I proposed that the Conference's resolution should also speak in favour of the establishment of IRENA. This proposal was adopted unanimously.

But opposition soon rose, for various reasons. Existing UN organizations that were partly active in the field of renewables, but with far less sweeping capability than what the IRENA initiative implied, spoke out against the establishment of the agency. Some members of the Organization of Petroleum Exporting Countries (OPEC) saw IRENA as a potential threat, opposing its establishment. The idea was also rejected by those that simply lacked the vision to see the potential for renewable energy sources to supply the world's energy needs. Finally and predictably, the conventional energy organizations resisted the emergence of a new and focused agency. Until now, even though the need for a renewable world vision has become so overwhelming no-one has been able to explain how the global spread of renewables can be carried without an appropriately empowered and newly chartered institution dedicated to the global renewable revolution, and matching the charters and impact of, say, an IAEA or IEA.

For many years, at international conferences in numerous countries, I have advocated the establishment of IRENA. Prerequisite for the founding has always been that one or more governments would take the initiative and build a coalition of like-minded countries. To avoid the notorious compromises and lack of a clear focus of past UN and other efforts, the focus was on purely intergovernmental alliances, an entirely new initiative free of historical shackles. One important recent milestone on the way towards establishing IRENA has been the 2004 International Parliamentary Forum on Renewable Energies, which was hosted by the German parliament, taking place in parallel to the governmental conference 'Renewables2004'. I convened 300 members of parliament from 70 countries to take part in this conference. The Final Resolution states:

> Promoting renewables requires new institutional measures in the field of international cooperation. To facilitate technology transfer on renewables and energy efficiency and to develop and promote policy strategies, the most important institutional measure is to establish an International Renewable Energy Agency…, which should be set up as an international intergovernmental organization. Membership would be voluntary, and all governments should have the opportunity to join at any time. The Agency's primary tasks would be to advise governments and international organizations on the development of policy and funding strategies for renewables use, to promote international non-commercial technology transfer, and to provide training and development. (www.ipf-renewables2004.de/IPF_Resolution_en.pdf)

On 26 and 27 January 2009, IRENA's founding conference and inaugural preparatory commission meetings were held in Bonn, with 75 inaugural signatories and more than 120 participating nations. This move was necessary and long overdue, for reasons that have now become plain and commonplace. On 29 and 30 June 2009 the agency's second preparatory meeting was held in Sharm-el-Sheikh where the number of members jumped to 136, as the US, Australia, Japan and Liechtenstein joined. The assembly chose the agency to be headquartered at Masdar City in Abu Dhabi, with Bonn and Vienna as specialised headquarter centres, and France's Hélène Pelosse serving as Interim Director General. IRENA's success will be measured by how effectively it pursues the goal of a fossil fuel and nuclear free world.

References

ALTER (Le Groupe de Bellevue) (1978) *A Study of a Long-Term Energy Future for France based on 100% Renewable Energies*, reprinted in *The Yearbook of Renewable Energies 1995* (1995), James and James, London

Federal Ministry for the Environment, Nature Conservation and Nuclear Safety (2008) *Founding an International Renewable Energy Agency (IRENA). Promoting Renewable Energy Worldwide*, Federal Ministry for the Environment, Nature Conservation and Nuclear Safety, Public Relations Division, Bonn, www.irena.org/downloads/IRENA_brochure_EN.pdf

Fischer, D. (1997) *History of the International Atomic Energy Agency: The First Forty Years*, Division of Publications, International Atomic Energy Agency, Vienna, www-pub.iaea.org/MTCD/publications/PDF/Pub1032_web.pdf

International Parliamentary Forum on Renewable Energies (2004) 'Renewable energies – The Challenge for the 21st century', www.ipf-renewables2004.de/IPF_Resolution_en.pdf

Kendall, H. and Nadis, S. J. (eds) (1980) *Energy Strategies: Towards a Solar Future. Report of the Union of Concerned Scientists*, Cambridge University Press, Cambridge

Parra, F. (2004) *Oil Politics, A Modern History of Petroleum*, I. B.Tauris, London and New York, www.amazon.com/Oil-Politics-Modern-History-Petroleum/dp/1860649777#reader

Chapter Three

The Renewable Imperative: Providing Climate Protection and Energy Security

Hans-Josef Fell

One world summit follows on the heel of another. Whether G8 Summit, European Council, World Climate Conference, East Asia Summit or the United Nations Framework Convention on Climate Change (UNFCCC) Conference of the Parties (COPs) – the energy conundrum has moved squarely into the centre of attention. Yet while some conferences focus on climate protection and others on energy security, no summit has produced a solution. Some seek to stabilize the climate by reducing emissions. Others still regard the intensified development of crude oil, natural gas and hard coal as key to ensuring energy security. A bizarre paradox has developed: by confusing energy security with conventional fuels – confounding ends and means – the latter group's aims are diametrically opposed to that of the former. There is only one solution that meets both climate protection and energy security objectives: the switch to renewable energies – completely, worldwide and within the shortest timeframe possible.

The paramount need to replace mineral oil, natural gas, coal and uranium with renewable resources is blatant, and yet many oil corporations and research institutions still maintain that no supply shortages will occur in the coming decades, pinning their hopes on offshore, deep-water, Arctic and non-conventional oil sources.

However, scientific studies, such as those from the Energy Watch Group (EWG) paint a worrying picture. Current studies conclude that global oil extraction passed its peak in 2006, when taking into account all estimated remaining production capacity.

In the coming years, oil extraction will decline by around 3 per cent annually. This will mean that in 2030 only around 50 per cent of today's level of 84 million barrels per day (mb/d) will be produced. The IEA in Paris is no longer capable of producing accurate oil forecasts. It claims in its *World*

Energy Outlook (WEO) 2008 that oil extraction could increase by nearly 30 per cent by 2030.

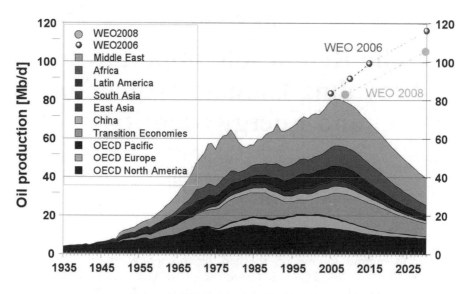

Figure 3.1 Oil production: World summary

Source: LBST/ Energy Watch Group
Note: Mb/d = million barrels per day

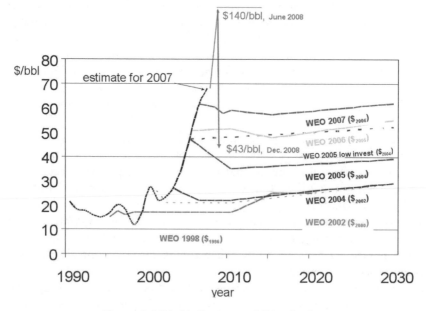

Figure 3.2 World oil prices and IEA price forecasts

Source: IEA
Note: bbl = billion barrels

The IEA's longstanding inability to estimate the world's resources is accurately portrayed in its oil price forecasts. Its forecasts had to be corrected downward for many years. For example, as recently as its World Energy Outlook of 2004, the IEA predicted a long-term oil price of under $30 per barrel in 2008. Yet by the middle of 2008 the oil price had approached $140. And still, most governments and industries rely on these incorrect forecasts. The rapid drop in oil price in the second part of 2008 resulted from the world economic crisis, itself triggered by a high oil price and the corollary, so-called financial crisis. But the oil price is bound to soon rise again, inexorably so, because of declining world oil production capacity.

Until very recently the IEA was not able to accurately assess the global fossil fuel resource condition. Precious time has been lost, in part because a widely trusted voice has refused to acknowledge that fossil fuels and nuclear power will not be able to meet the world's energy demand in the coming years. Those who rely on such announcements and continue to refuse to invest heavily in renewables and energy savings will contribute to a deeper global economic crisis than experienced today.

Just as the global oil supply is beginning to evaporate, so it is inevitable that the natural gas supply will soon follow suit. Natural gas is a very limited resource that cannot replace oil-based energy production. Coal-based energy production, as the most environmentally hazardous, should be abandoned as quickly as possible.

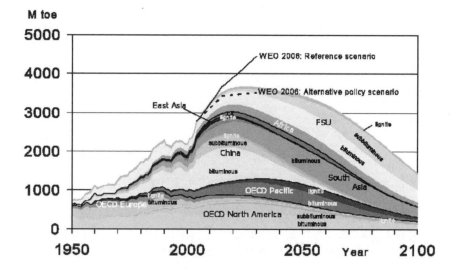

Figure 3.3 Projected coal production capacity in the coming years

Source: LBST/Energy Watch Group
Note: mtoe = million ton oil equivalent

The common assumption that coal will be readily available for the next 200 years is false. The EWG has verified that within the next few decades there will already be shortfalls in coal supply. Often the point is raised that coal cannot be affordably replaced by some emerging economies such as China, or that the world's largest coal exporters, such as Australia, have a legitimate interest in its continuing use of coal. The dream of carbon-free coal-fired power plants is not only held up by many proponents with a serious face, but increasingly pursued at a well-funded scale in the form of various carbon capture and storage (CCS) technologies. In Germany, a country that supports CCS with high investments from public research funds, an extensive study of its potential was published in March 2008 by the independent scientific consultancy office of the German Bundestag (TAB). The message is clear: CCS raises many scientific questions but no real answers; to date, there are no plants operating and all experiments appear to have failed. Also, electricity production with CCS requires about 30 per cent more coal. Given growing resource scarcity and inexorably rising pressures on world market prices, coal-fired power plants with CCS do not stand a chance against renewable energies, unless they are heavily subsidized. CCS is not likely to be practically available before 2020. By then its chances will look even worse than today.

The coal price has risen rapidly recently and demonstrates the arrival of an age in which scarcities become the norm, rather than the exception.

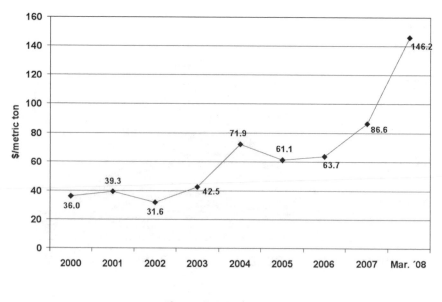

Figure 3.4 Coal prices

Source: Verein Deutsche Kohleimporteure e.V
Note: refers to MCIS steam coal marker price; first price calculated each month

Finally, uranium is the most limited conventional power resource of all. It is irrational and inappropriate to pursue nuclear power as a path forward, however strenuously pursued by some industry lobbies today. EWG shows that shortages in the supply of uranium required across the world's 439 reactors can be expected in the next few years, even though nuclear power supplies only 2.2 per cent of the world's energy consumption.

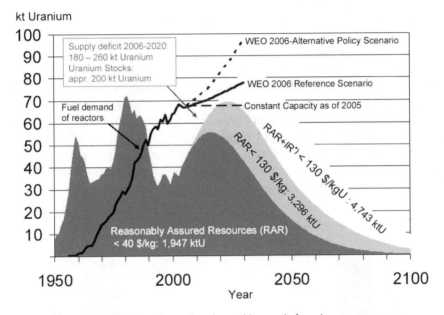

Figure 3.5 Uranium demand and possible supply from known resources

Source: LBST/Energy Watch Group
Note: IR* = inferred resources; kt = kiloton

South Africa provides an example of a failed power policy based on nuclear and coal energy. In 2007, a national emergency was declared because of power blackouts due to worldwide coal production and delivery shortages, with nuclear capacity being rigid and expensive and unable to compensate for the rising bottlenecks. Wisely, South Africa has since decided against building a new nuclear generator.

China closed a number of coal-fired power plants in summer 2008 due to coal shortage. In contrast, wind energy has quickly developed, and by the end of 2008, there were 67 wind turbine manufacturers in China (www.sonnenseite.com/index.php?pageID=80&article:oid=a11581&template =article_detail.html) China moves increasingly towards renewable energies.

Today we are already witnessing the first climatic disasters as a consequence of a rise in global temperatures approaching only 0.8°C – and some are still proposing to actually work toward 'limiting' the rise to 2°C. This position is incomprehensible, even if a reduction of emissions were to be actually possible, carbon dioxide and other greenhouse gases in the atmosphere are already exceeding stable levels. Any further emissions will bring about an additional increase in the Earth's temperature. The use of oil, gas and coal contributes around 80 per cent of all CO_2 to the atmosphere. Thus the genuine prevention of climate change demands the end to their use. Saving energy is a help but is not sufficient in itself. Fossil resources have to be substituted by renewables.

Climate problems can be solved only by two strategies: first, stopping greenhouse gas emissions (not just reducing emissions) by promoting zero emission technologies and completely stopping the use of fossil fuel and nuclear energies; and second, taking carbon out of the atmosphere, for example, by large-scale forestation projects and storing carbon in the ground as humus. The new hydrothermal carbon biochar technology (HCT) has huge potential. Markus Antonietti from Potsdam's Max Planck Institute in Germany hopes that this technology could reduce the worlds CO_2 concentration from today's 387ppm to 350ppm within some decades. HTC derives energy from plants and removes carbon from the atmosphere and stores it in the soil. Biochar also helps restore degraded soil into fertile land.

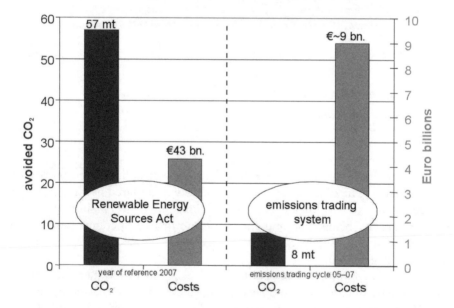

Figure 3.6 FiT under the Renewable Energy Sources Act versus emissions trading

Source: www.energie-verstehen.de/Energieportal/Navigation/energiemix,did=249684.html; Zugriff 27.11.08, BMU-Publikation 'Erneuerbare Energien in Zahlen: nationale und internationale Entwicklung', KI III 1, Stand Juni 2008; Antwort der Bundesregierung auf schriftliche Frage zu Emissionshandel und CO_2-Einsparung vom 03.04.2008, Schlemmermeier, Schwintowski: ZNER Jg. 10/3/2006, Seite 195
Note: mt = millions of tons

A policy for climate protection and energy security can only be achieved through a clear, consistent and uncompromising policy for renewable energy resources. In particular, this means: FiT laws and tax exemption for renewables; an end to subsidies for fossil and atomic energy; campaigns for research, development and education in renewable energy resources; and an end to legal resistance during the approval process.

My doubts concerning emissions trading are growing. In Germany it has not been very successful so far. While laws for renewable energies have saved more than 120 million tons of CO_2 of annual production, with an added cost of only €4 billion, emissions trading has contributed only 9 million tons. But emissions trading generated €5–10 billion of unjustified additional profit for conventional energy companies.

Generally speaking, renewable energies bring only advantages and no burdens. They create jobs due to economic development and reduce prices for energy by creating independence from rising fossil resource prices. They bring a secure energy supply and cancel out the motivation for wars over oil.

But can we really afford to abandon the use of crude oil, natural gas, coal and uranium? There is a clear answer to this question. Precisely because the global economy is so profoundly dependent on fossil resources, it must develop another resource base very quickly. Otherwise, the increasing depletion of oil resources, the drastic rise in oil prices and the simultaneous escalation of the global environmental crisis will precipitate an unprecedented worldwide economic crisis. The current crisis is in part a result of high oil prices in the summer of 2008.

It is possible to make a fast total switch away from crude oil, natural gas, coal and uranium. The natural supply of renewable sources of energy – solar, wind, hydroelectric and geothermal power, bioenergy and ocean energies –

Sun – the energy of the 21st century

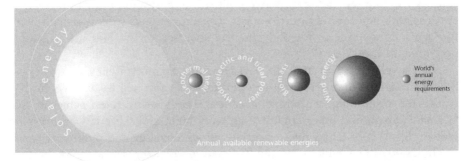

Source: FVS/DLR

Figure 3.7 The supply of renewable energy

Source: German Solar Power Research Association

offers many thousand times' the world's entire annual energy requirement. Enough solar radiation, in particular, reaches the Earth to meet today world's energy demand 15,000 times over (www.fv-sonnenenergie.de/fileadmin/bildarchiv/grafiken_und_charts/Erlaeuterung_zu_Grafik_EE-Potenziale.pdf).

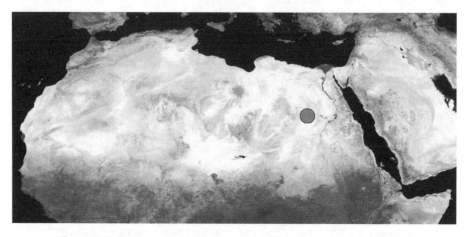

Figure 3.8 Concentrating solar power potential versus electricity demand

Source: NASA

Note: 1 per cent (•) of the Sahara's surface is enough to meet the world's entire electricity demand using CSP technologies

It is possible to achieve the goal of 100 per cent renewables worldwide. Studies have long shown that entire regions could meet their own energy demand with self-produced energy from renewable energy sources. If we heed the principles laid out in the Energy Rich Japan study by ISUSI (www.energyrichjapan.info/pdf/ERJ_fullreport.pdf), we can come to produce our electricity, heating, cooling and transport fuels from renewable energy sources within a few decades. Active political support and policy frameworks are crucial to build further development. The expansion is still restricted to far too few nations. Successful and efficient laws for renewables are urgently needed in all countries.

In the electricity sector, the German and Spanish Renewable Energy Sources Acts are among the most important and most successful laws in the world for promoting renewable energies. The growth rates achieved by Germany, Spain and other countries are very high, and there has been rapid industrialization of renewables, especially of wind energy and photovoltaics.

In 2000, the members of the German parliament set a target in the Renewable Energy Sources Act for 12.5 per cent of electricity to come from renewable sources by 2010. We were told that this target was unrealistic and unachievable. And yet by the end of 2008 a 16 per cent share had already been achieved. This shows that renewables can grow much faster than is often assumed. What is crucial is the political framework, such as FiT. Worldwide, renewable electricity is increasing at such a rate that worldwide energy demand could be met by renewables within a few decades.

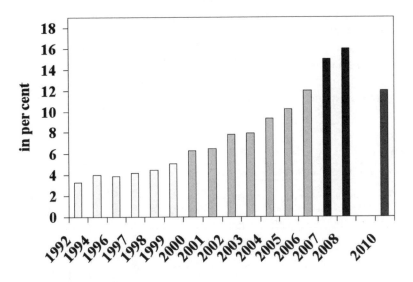

Figure 3.9 Share of renewables in gross electricity consumption in Germany

Note: right side of graph shows the target set by government in 2000 for 2010

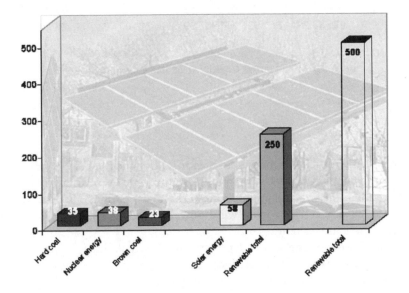

Figure 3.10 Renewable energies as an engine for job creation in Germany

Source: www.bee-ev.de/uploads/HG_Energieluecke_080423.pdf
Note: job numbers in thousands for 2007 with the final bar (500) being a prediction for 2020

Evidence for the development of renewable energies is provided by the rapid increase of jobs in this industrial sector over the last few years. In 1998, only 30,000 persons were employed in the renewable energies industry in Germany, just as many as in the nuclear industry. By the end of 2007, the number of persons working in the renewable energies sector had increased to 250,000. Experts expect that around 500,000 renewable industry jobs will be newly created by 2020 in Germany.

The most attractive policy framework is created by FiTs, which have proved to be remarkably effective in the promotion of renewable energy in Germany and Spain. A look at laws in other countries reveals that the German FiT law is the most successful instrument of all.

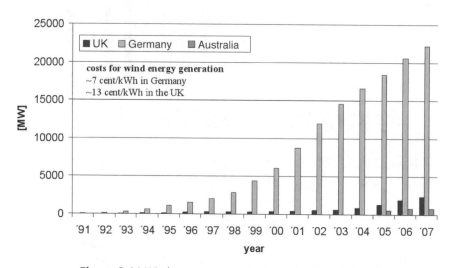

Figure 3.11 Wind power capacity in Germany, UK and Australia

Source: BSW, EPIA, ISES
Note: kWh = kilowatt hour

The UK failed to reach Germany's level of installed wind capacity, despite the fact that it is windier in the UK than in Germany (true even when discounting the difference in the size and populations of these two nations). In addition, the cost of a kWh of electricity from wind power in the UK, at $0.13, is almost twice as high as in Germany.

A comparison of the rates of growth in photovoltaics in Germany and Japan also provides persuasive evidence that FiTs represent a more successful industrial policy than the state subsidies or quotas and certificate laws that are used in Japan. The initial successes of the photovoltaics industry in Japan have been far surpassed by Germany since the Renewable Energy Sources Act was passed.

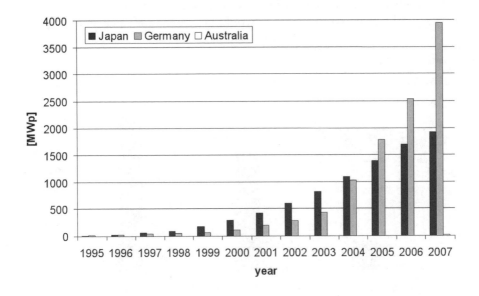

Figure 3.12 Photovoltaics capacity in Germany, Australia and Japan

There will be further great technological development in all sectors: wind power, solar energy, geothermal energy and especially marine energy. The energetic use of oceans is still to be made accessible. Many countries currently locked into a fossil fuel trap could rapidly become market leaders for wave, stream and tidal energy in the context of sound FiT laws. Australia is a good example.

A successful FiT law has components that include privileged grid access, entitling investors to connect and feed into the grid. The tariff must be paid for over an extended period of at least 20 years. This provides the necessary security for investment. The feed-in tariff must also be high enough for investment to be profitable. The cost of the FiT should be reflected in the electricity price, and there should be no cap of the total amount of power generated. Finally, there should be a guaranteed feed-in period and no obstacles should be erected in the form of approval procedures.

The introduction of the Renewable Energy Law in Germany continues to be controversial. Yet counterarguments of any substance have been lacking. The initially high cost of renewables is bound to fall in the coming years, allowing them to be carried in the economy of scale. Similarly, the law does not concern subsidies because tax funds are not at all involved. The entire financing of the Renewable Energy Law is through private capital, without public funds. The state only specifies the framework for how investments in renewables are able to become economically profitable. The market alone covers the remainder.

The introduction cost of renewables is also minimal. The additional cost for a typical household in Germany would be less than €3 per month.

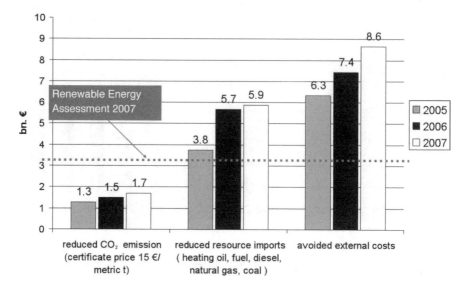

Figure 3.13 Costs avoided due to renewable energy

Source: BEE

All in all the performance of the national economy in Germany has been improved by renewable energies. While renewables accounted for €3.5 billion in additional costs per year, over €15 billion could be saved due to reduced imports of fossil fuels, avoided external environmental damage and reduced costs for emission certificates.

The most significant advantages of FiTs for renewable energy resources, especially for wind, biogas, photovoltaic, hydro and geothermal, support their worldwide introduction. Although the advantages of the FiT for renewables are well-known, there is still much resistance. The conventional fossil and atomic energy companies practise widespread lobbying for their own interests and often against renewables.

The Renewable Energy Act was introduced in Germany by the Green Party and Social Democrats in 2000, against the votes of the Conservatives. But now the conservatives in the German parliament accept the FiTs for electricity from renewable energies and even consider them to be indispensable. Until recently, this situation was unimaginable: during the national election campaign in 2005, the Conservatives demanded the abolition of FiTs, but the extremely successful development of electricity production from renewable energies has come to be one of the driving forces behind recent economic growth in Germany.

The heating and cooling sector is also very important for renewables. As far as heat supply is concerned, there is no development of renewable energies comparable to the electricity sector in Germany. For years, the development rate has been stagnating because a renewable heat act is still missing in Germany. The parliamentary group of the Green Party has made proposals for the framework of an effective heat energy act. The recommendations of the European Commission in January 2008 are suitable for a rapid development of renewable heat in the European Union. The proposal from Greens and the EU-Commission is to make renewable investment in heating and cooling obligatory by law, in both new and refurbished buildings. A change towards renewables in the transport sector is also important and indeed possible. There are two important dimensions to this shift: improved biofuels and electric engines powered by green electricity.

Biofuels are often erroneously blamed for rising food prices, but such rises are caused mainly by increasing oil price, bad harvests caused by climate change and the huge meat consumption. Land use for meat production is five to ten times as intensive as land use for plant-based nutrition. Still, the sustainable cultivation of biofuels is necessary. Disregarding social and ecological farming standards can indeed aggravate food supply problems. Intensive land use often results in the degradation of soil. Organic farming and biochar, however, improve soil fertility and support climate protection. Mixed cropping and agroforestry are examples of ecological agriculture with ample harvests.

The development of chemical products based on renewable resources is also very important for climate protection. Bioplastics solve many difficulties such as resource scarcity, climate change and waste problems. However, genetic engineering is no solution because it creates further ecological and social challenges. Genetic engineering is simply not necessary either for nutrition and biofuels or for bioplastics.

I have been the proud owner of a solar car for many years. The amount of electricity it consumes is small and can be produced by photovoltaic solar parks. I only need 10 square meters of photovoltaics on the roof of my home to enable my car to run 10,000 kilometres per year. What is necessary is to convert automobiles to emission-free drive systems powered by electricity from renewable energy sources. The most important strategy to introduce renewables into the transport sector is to convert the focus of car producers from combustion machines to electricity cars, buses, cycles and others. But they must run only with renewable electricity.

Education and the transfer of knowledge are essential factors to meet the challenge of transforming our energy system. The successful launch of IRENA will greatly aid in this important task.

Chapter Four

100% is Possible Now

Harry Lehmann and Stefan Peter

Today's world economy does not use the Earth's resources in a sustainable way. Humanity is rapidly coming closer to the exhaustion of conventional and nuclear finite energy reserves. Now it is widely accepted that the mid-depletion point of oil reserves has already been passed in non-OPEC countries, and will be passed in the OPEC countries within the next decades (according to EWG (2008a), peak oil, i.e. global maximum oil production capacity, was passed in 2006. Additionally EWG studies on global coal and uranium resources showed that these will also become scarce in the near future. Last but not least, the environment faces increasing threats, ranging from the well-known problem of human-caused climate change to the erosion of fertile soil, water pollution and the various effects of manmade toxins and nuclear waste.

A sustainable energy supply has to be fully based (100 per cent) on renewable energies and must use available resources most efficiently. Today, at the advent of the industrial solar age, there are eight basic technologies in the market (biomass, hydropower, wind energy, solar thermal collectors, photovoltaic systems, centralized solar thermal power plants, solar architecture and geothermal energy) and a number of experimental technologies are raring to go. These technologies tap renewable energy resources in a magnitude that exceeds current global energy consumption by many times, and can provide energy for millions of years.

A solar energy system to supply energy reliably throughout the year includes the consistent use of local renewable energy sources wherever possible. A second basic requirement of such a fully renewable energy supply structure is the intelligent interchange of energy between regions. This interchange can be managed by the power grid, gas networks (using solar-generated hydrogen) or by transporting biomass. National or international networks can be used for system balancing, for example, to transport regional surpluses to regions with deficits or to storage facilities. In this chapter, we focus on regions where an electrical infrastructure already has been installed, but renewable energy technologies are as well able to supply the needed energy down to the size of a single house or a small village.

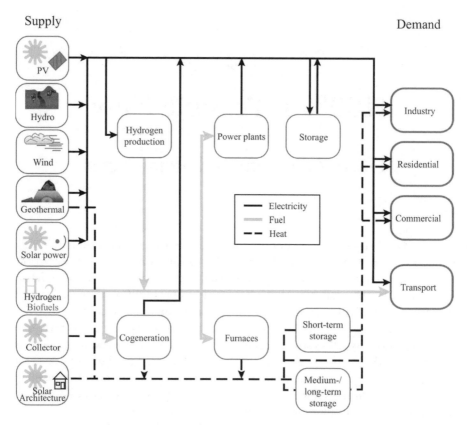

Figure 4.1 The structure of a renewable energy system

Source: ERJ (2003)

Different renewable energy technologies as well as a diversified regional distribution, combining several regions with varying strengths and weaknesses, will mutually replenish a functioning energy supply system throughout the year. This approach helps balance the fluctuations in energy provision that can occur with some renewable energy technologies (such as wind and photovoltaic). If, for example, the wind stops blowing in one region, power can directly be supplied by surpluses in other regions or, if this does not suffice, other regional sources, such as local biomass power plants, or even plants in other regions can deliver the required power.

Such an energy supply structure requires much more intelligent management than today's energy supply. This starts with regulating the system, which includes the planning of energy production with the help of weather forecasts, and ends with managed consumers, adapting their energy needs to the availability of electric power. Cogeneration units (motor-driven cogeneration units of all sizes are already available on the market, fuel cells or renewable powered gas turbines will follow in the coming years) deliver heating/cooling and power by burning solar hydrogen, biogas or biomass and are part of a new kind of intelligent consumption/production. Consumers can then provide themselves with heating and electricity, and even more, these 'personal power plants' can also deliver power to the grid if, for example, a central control authority needs to allocate a peak load or they can create a virtual power plant by connecting several personal power plants with windmills, photovoltaic and other sources, delivering electricity on demand. Management and storage of electrical surpluses (whether in the form of electricity, fuel (hydrogen), heat or cold) is viable for a 100 per cent renewable energy supply. And even parking electric cars can contribute to system stability and balancing, as they then offer both, storage and power on demand.

Foresighted management can guarantee a stable energy supply for consumers by combining those technologies with variable or seasonal energy production and those whose energy sources are available at any time in an (inter)national exchange structure. Modern computers and the communication technologies that gave rise to the internet offer this possibility today.

Systems and approaches as described above are often investigated in scenarios. Scenarios are a look ahead, from the present into the future, and allow us to study future development under specific surrounding conditions. Many scenarios and studies dealing with the future of energy supply have been prepared and published in recent decades (see Le Groupe de Bellevue, 1978, Nakicenovic and Messner, 1982; and www.solarmissionpossible.info).

Europe

One study investigating the possibility of such an energy system is the 'Long-Term Integration of Renewable Energies into the European Energy System' (LTI). The LTI project will work on 'extreme' scenarios with very different but ambitious economic, social and ecological goals over the next decades (LTI, 1998).

Based on two simplified archetypes of behaviour – exhibited by those who are motivated to protect the environment ('sustainable' scenario) and those who are interested in consuming ('fair market scenario') – two scenarios were developed that result in an 80 per cent reduction of CO_2 by 2050. Because of the extremely varying assumptions, the examined scenarios represent two extremes of possible development and are not meant as a prognosis for the future. Rather they were designed to learn as much as possible about supplying solar energy to Europe. The reality will be a mixture of different trends and will incorporate aspects of both scenarios. A third optimized version of the supply system is shown in the 'opti' scenario – a 100 per cent renewable energy scenario (Lehmann, 1998). This last scenario is much better in terms of availability of electricity over the year.

The LTI project shows that the European energy system can be changed until 2050 to use energy in a sustainable way. There are no fundamental technical or financial hurdles that inhibit an exclusively solar/renewable energy supply system for Europe.

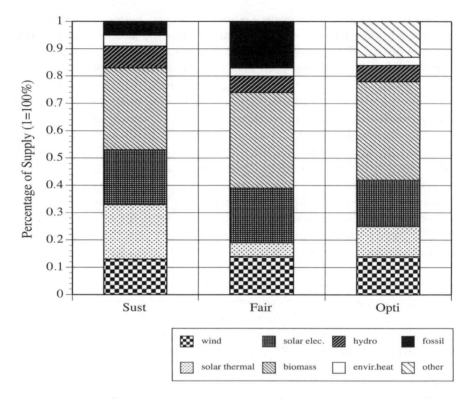

Figure 4.2 Share of energy sources providing energy for Europe in 2050 in three different scenarios

Source: LTI (1998)

Note: Energy efficiency and savings have lowered demand by 38–62 per cent. Geothermal energy, imported solar-generated hydrogen and others are not included in the sustainable and fair market scenarios and are together in the opti scenario under 'other'

Germany

In February 2000, the German Bundestag constituted the Enquete Commission on 'Sustainable Energy Supply Against the Background of Globalisation and Liberalisation' (Enquete, 2002). The Commission had the mandate to provide scientific evidence as a basis for the German Bundestag's future decision-making in the field of energy policy. The consensus view of the Commission was that Germany's energy supply system was not sustainable at that time. In order to assess the prospects of sustainable development up to the year 2050, the Commission examined economic and technological capabilities as well as options for practical and political action.

Based on 14 developed scenarios and the evaluation of additional studies, the Commission's conclusion was that it is technically feasible and economically possible to reduce greenhouse gas emissions by 80 per cent in a modern industrialized country, even if nuclear energy is phased out, by shifting to a sustainable energy system, based on renewable energy resources and efficient energy technologies. The Commission's work also outlined that it is possible to cover the total energy demand by means of solar/renewable energy sources.

In the medium and long term, a restructured energy system will not be much more expensive than the present one and will even create more jobs than a conventional/fossil system does. Initially, higher investments will be needed to push this development. In 2050 the additional costs compared with a 'business as usual' scenario amount to €15 per capita a month in the year 2050 (80 per cent reduction of climate gases).

Germany will take a lead in climate protection and supports a 30 per cent reduction by the EU, with its own commitment of a 40 per cent reduction by 2020 compared to 1990 levels. To achieve this target, the German government has adopted an extensive set of measures (Integrated Energy and Climate Programme – IECP). In a scenario developed by the German Federal Environment Agency (UBA), the feasibility of achieving this target was demonstrated. The most important measures are increasing the efficiencies of energy use (for example, in the housing sector) and the introduction of renewable energy technologies.

Japan

Another study is the 'Energy Rich Japan' Report (ERJ, 2003), which provides an analysis of Japan's current energy demand across the industrial, residential, commercial and transport sectors (using 1999 data) and shows how energy demand can be reduced substantially in all four sectors by adopting most energy efficient technologies. The study shows a possible halving of Japan's end-energy use from almost 15,200 petajoules (PJ) in 1999 to a level below 7500PJ.

To meet this new reduced demand, the report includes a renewable energy supply model to supply electricity, heat and fuels. The ERJ Report produced six renewable energy scenarios, all of them providing 100 per cent renewable energy for Japan. The starting point is a basic model (Scenario One), which provides more than 50 per cent of total energy needs from domestic sources of renewable energy in Japan, including fuels for transport. Each subsequent scenario provides variations or expansions on Scenario One, gradually reducing the reliance on imported energy, factoring in different population projections and expected improvements in renewable generation capacity and energy efficiencies, until by Scenarios Five and Six, energy imports are not required any more.

To ensure that renewable technologies get used to their best advantage and to guarantee supply security, the report includes a computer simulation of the Japanese supply system with a temporal resolution of a quarter hour. The system uses hourly resolved and detailed meteorological data from 153 sites around Japan, which provide information on renewable sources of wind, solar radiation and temperature to reflect the changing weather conditions for power calculation.

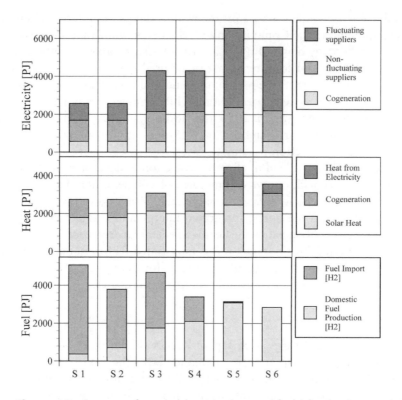

Figure 4.3a Structure of supply (electricity, heat and fuels) for the six scenarios

Source: ERJ (2003)

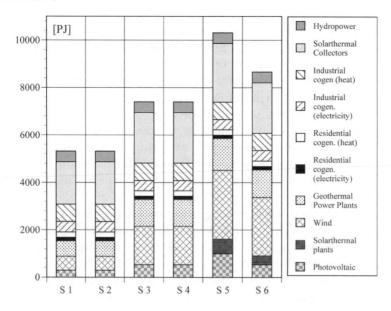

Figure 4.3b Gross energy production from different technologies in the six scenarios

Source: ERJ (2003)

The challenge in designing a reliable, fully renewable energy system was to find a combination of technologies where the pros of some types balance out the cons of the others. Reserve capacities are necessary as a backup for fluctuating sources, especially in the electrical system, but they can be minimized by designing a combination of renewable technologies where fluctuations in production match a varying demand as much as possible. Fluctuating sources, such as wind and solar, were combined with adjustable 'supply on demand' sources such as geothermal plants and hydropower to guarantee a reliable supply of energy throughout the year from domestic Japanese energy sources, regardless of seasonal or daily variations. Surpluses in the electrical supply system were stored as hydrogen (for later use in various types of thermal plants) or in pumped water storage systems, both giving the opportunity to deliver energy on demand at times when fluctuating sources do not produce enough. Thus the scenarios showed that a complete energy supply of Japan from renewable sources is possible with a domestic share of up to 100 per cent.

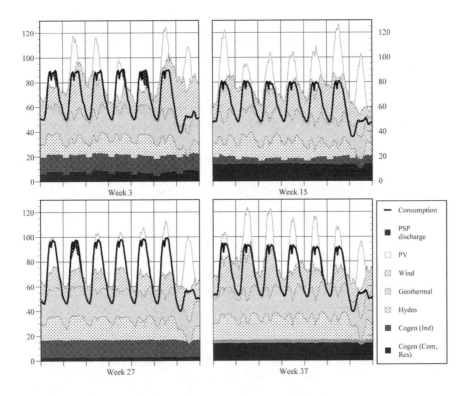

Figure 4.4 Simulation dynamics for four exemplary weeks, showing the contributions of all sources and energy demand

Source: ERJ (2003)

Catalonia

The objective of initiating the Catalonia study (SolCat, 2007) was to show Catalonia's ability to supply its own need for energy from renewable sources and, thus, to provide a fact-based vision of a future energy supply as an alternative to the present fossil/nuclear system. The study is focused on Catalonia's electric energy demand – and how it can be reduced – and the design of a reliable renewable electricity supply system.

The scenarios highlight a development towards halving electricity intensity in the three most important sectors of electricity consumption until 2050, which, of course, is a great challenge, but feasible from a technological point of view.

The future development of generating capacities for all renewable technologies was calculated using so-called 'logistic growth functions', showing the typical s-curved shape for growth with saturation effects in the later stage of development. This reflects the underlying assumption that growth cannot be unlimited if any of the resources growth depends on is limited. This approach also required incorporating assumptions regarding the future development of technology specific investment costs. While wind energy, geothermal power plants, biomass plants and solar thermal power plants were expected to show half of today's specific investment costs by 2050, the specific costs of photovoltaics were expected to fall to one third of today's costs; hydropower was assumed to remain on current cost levels.

The 'Fast Exit Scenario' shows an increase of renewable generating capacities from less than 200MW in 2006 to about 6400MW in 2030 and further to almost 12,000MW in 2050. Wind energy contributes most to the total renewable capacity – almost three fourths (2030) respectively two thirds (2050) of the total renewable capacity consists of wind energy. Photovoltaic shows a dynamic extension too, resulting in an 11 per cent share of total renewable capacity in 2030 and 23 per cent in 2050. The shares of biomass, geothermal, solar thermal plants and additional hydropower are substantially lower if compared to wind energy or photovoltaic.

Validation of supply security was based on dynamic computer simulation of electricity supply for four representative weeks for all the four seasons of the year. The simulation showed no indication of undersupply at any time.

The two scenarios of the Catalonia study show the feasibility to achieve a fully renewable electricity supply, one until 2035 (Fast Exit Scenario), the other until 2045 (Climate Protection Scenario). The realization of these goals is not a matter of potential, but it is a matter of setting and pursuing ambitious targets, encouraging policy and people and – of course – the financial investments Catalonia and its people are willing to take. The scenarios show that the financial aspect is not as big an obstacle as one might expect. With an annual investment into renewable capacities peaking at €104 (at 2006 value) per inhabitant in the 'Fast Exit Scenario' (2050) and €85/capita in the 'Climate Protection Scenario'.

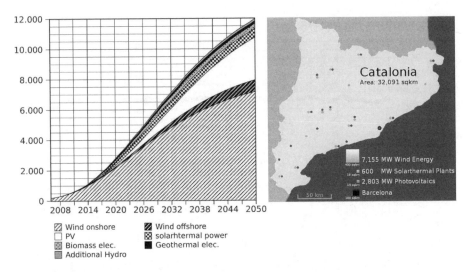

Figure 4.5 Areas required to install sufficient renewable energy generating capacity to supply 100 per cent of electricity demand in Catalonia (right) and development of renewable electricity generating capacities in the 'Fast Exit Scenario' (left)

Source: SolCat (2007)

Note: On the right, wind energy, the renewable energy technology with most installed capacity, would require about four times the area of Catalonia's capital Barcelona. PV and CPS both require about the same area for installation, which is about one fifth of the capital's area for each of these technologies

Compared to the Catalonian gross domestic product (GDP) (€181,029 million in 2005), the annual costs of the scenarios are 0.2 per cent of GDP for the 'Climate Protection Scenario' and 0.3 per cent for the 'Fast Exit Scenario' on average.

Renewable Energies World Outlook 2030

The Renewable Energies Outlook 2030 scenarios (EWG, 2008b) differ from the above described scenarios, as they do not outline a 100 per cent renewable supply. Rather the 2030 scenarios deal with the financial-driven global extension of renewable capacities up to 2030. Growth was calculated with regard to global renewable potentials (using the logistic growth approach too) and reduction of specific costs due to the massive extension of production capacities. Two scenarios, a 'low variant' and a 'high variant', assume different investment figures, defined as 'investment paths' with successive increasing annual investments towards 2030.

Presuming strong political support and a barrier-free market entrance, the dominating stimulus for extending the generation capacities of renewable technologies is the amount of money invested. One basic assumption within the Renewable Energies Outlook scenarios was a growing 'willingness to pay' for clean, secure and sustainable energy supply, starting with a low amount in

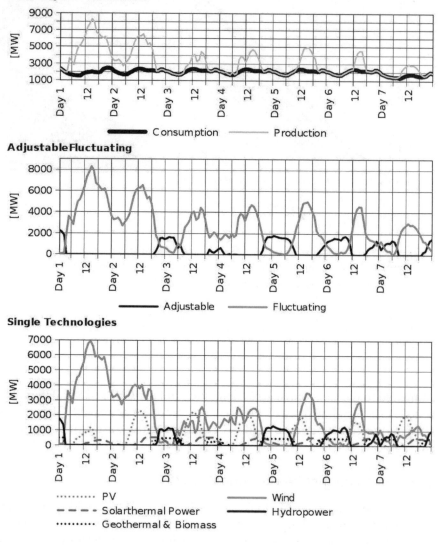

Figure 4.6a Electricity demand and supply in spring

Source: SolCat (2007)
Note: Production peaks occur around midday, driven by good solar radiation and wind conditions. At no time does demand exceeds production. Due to the good performance of solar and wind energy, fluctuating suppliers contribute most to electricity supply, with only minor contributions from adjustable suppliers (hydropower, geothermal and biomass).

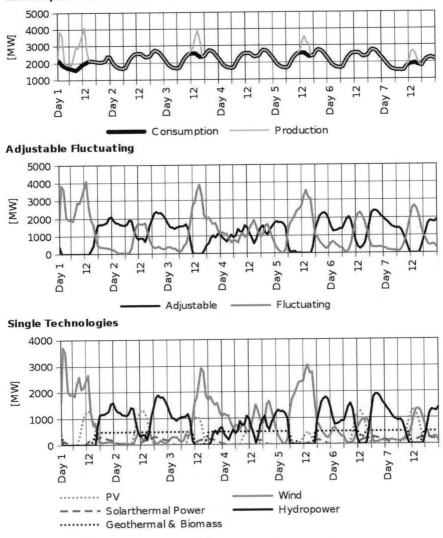

Figure 4.6b Electricity demand and supply in autumn (November)

Source: SolCat (2007)
Note: Electricity demand is always met, with peaks of energy surplus showing, of a few hours per week. Looking at the contributions of the single technologies (third graph) shows that solar radiation has considerably dropped since summer and that wind performance is comparably weak during the simulated week, with almost no wind power on days two and three and peak production at about 1000MW on days six and seven. Consequently hydropower, geothermal and biomass have to make great contributions to satisfy electricity demand.

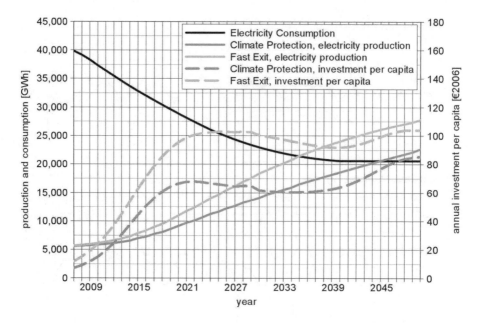

Figure 4.7 Development of renewable energy production and consumption:
Development of annual investments per capita

Source: SolCat (2007)

2010, which then successively increases towards 2030. Finally, specific target levels for annual investments per inhabitant (capita) get reached by the year 2030. The targeted amounts differ for the various regions of the world (see Table 4.1). In global average, €124 are spent in 2030 per capita in the 'high variant'. In the 'low variant' the target for 2030 is half that amount (€62 per capita and year).

Absolute investments in 2030 are approximately €510 billion in the 'low variant scenario' and about €1021 billion in the 'high variant'. The biggest single investor in both scenarios is China, followed by South Asia – both regions having a high percentage of the world population – and OECD Europe, which is less populated but shows considerably higher spending per inhabitant in 2030. OECD Pacific has the lowest investment figure, behind Africa, the Middle East and Latin America.

To provide a better feeling for what such investment figures really mean with regard to today's real world, Figure 4.9 compares the renewable investments of the Renewable Energies Outlook 2030 study to global military expenditures in 2005. Only the 'high variant' shows renewable per capita investments coming close to the military expenditures of 2005. Another illustrative comparison is the amount of money spent by each German in 2005 on culture-related activities – which is of the magnitude of €100 annually.

Table 4.1 Targets for annual investments into renewable generating capacities in 2030 in the Renewable Energies Outlook 2030 'low variant' and 'high variant' scenarios

Region	Investment per capita per year in 2030 (€2006/capita/year)		Total investment budgets in 2030 (€2006 billions)	
	Low variant	High variant	Low variant	High variant
OECD Europe	111	223	60	121
OECD North America	110	220	59	118
OECD Pacific	112	224	22	44
Transition economies	91	180	31	60
China	102	204	149	299
East Asia	41	81	33	66
South Asia	35	71	73	147
Latin America	46	91	26	52
Africa	20	41	30	59
Middle East	101	202	28	55
Global scale				

Source: EWG (2008b)

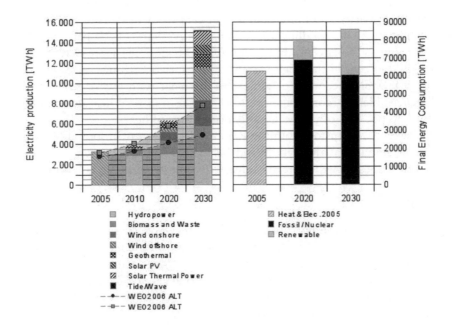

Figure 4.8 Final electricity and heat demand and renewable shares until 2030 in the 'high variant' scenario

Source: EWG (2008b); IEA (2007)

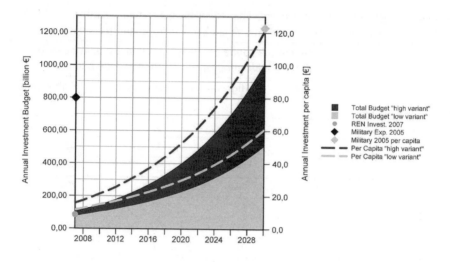

Figure 4.9 Development of investments in the Renewable Energy Outlook 2030 scenarios and military expenditures as of 2005

Source: EWG (2008b)

The development that could be initiated by investments as assumed in the 'high variant' is impressive: the OECD region will be able to cover more than 54 per cent of its electricity and more than 13 per cent of its heat requirements from renewables in 2030, totalling to a final energy share of 27 per cent ('low variant': almost 17 per cent). In the non-OECD region, the share of renewables rises to 30 per cent in the 'high variant' ('low variant': 18 per cent). Increases due to renewables account for almost 68 per cent with regard to electricity, while renewable heat contributes about 17 per cent of final heat demand ('low variant': 36 per cent of electricity and 11 per cent of heat).

Drawing a conclusion from the Renewable Energies Outlook 2030 study, it must be stated that the widespread belief that to change energy supply to renewable resources would be an unfeasible economic burden is wrong.

Conclusion

Future energy needs can be fully supplied by renewable energies. The scenarios described here have never been disproved but were considered utopian, although they relied on the current status quo of the technologies and projected the introduction of these renewable energies on to total energy demands.

The analysis of scenarios that consistently go as far as a 100 per cent supply from renewable energy shows the following:

- A reduction of greenhouse gas emissions by 50 per cent worldwide and by 80 per cent in industrialized countries by 2050 (compared to 1990 levels) is technically and economically feasible from today's perspective. The path towards a full renewables-based and efficient energy system is a realistic future option and not a dead end. There is no need for nuclear power, neither fission nor fusion.

- The supply and demand systems described in the scenarios offer possible target ranges for restructuring the energy sector. Any restructuring towards renewable energies will not need to be limited to the ideas described in those reports. Other systems and other combinations of technologies are also possible.
- Since supply security in the electricity sector is crucial – production and consumption must match at all times – some of the energy systems were reproduced in computer-based simulations. The simulations showed that a system totally based on solar energy, with suitable design and control, works reliably all year round.
- Although investments will initially be necessary to stimulate development towards a sustainable energy system, the restructured system will not result in higher costs than the present system in the medium and long term. All these scenarios are economically feasible.

Analysing older scenarios, it is obvious that the market introduction and expansion of renewables have beaten even the expectations of optimistic scientists in the regions that implemented suitable support mechanisms. 'First-mover' countries (such as Germany) are in a highly favourable win–win situation, due to the creation of jobs and export opportunities. In Germany 250,000 new 'clean tech' jobs have been created in recent decades in the field of renewable energy technologies. In nine years, Germany more than tripled the amount of renewables in the electricity market and today renewables cover 15 per cent of supply. This should make us confident that these scenarios can in fact be implemented.

Today's society must take action to implement a renewable strategy. The most important step is to start right now, since every day that passes by without enforcing a renewable energy strategy only increases and complicates the problem because energy consumption is increasing, money is still being invested in fossil/nuclear systems and finding ways to solve the problem of climate change merely gets postponed.

References

Enquete (2002) *Sustainable Energy Supply Against the Background of Globalisation and Liberalisation*, Enquete Commission, Deutscher Bundestag, www.bundestag.de/gremien/ener/index.html

ERJ (Energy Rich Japan) (2003) *Energy Rich Japan – A Vision for the Future*, www.energyrichjapan.info

EWG (Energy Watch Group) (2008a) Crude Oil: The Supply Outlook, Energy Watch Group, www.energywatchgroup.org

EWG (2008b) *Renewable Energy Outlook 2030 – Energy Watch Group Global Renewable Energy Scenarios*, www.energywatchgroup.org/Studien.24+M5d637b1e38d.0.html

IEA (International Energy Agency) (2007), *World Energy Outlook 2007*, IEA, www.worldenergyoutlook.org/

Le Groupe de Bellevue, ALTER (1978) *A Study of a Long-Term Energy Future for France Based on 100% Renewable Energies*, reprinted in *The Yearbook of Renewable Energies 1995* (1995), James and James, London and Le Groupe de Bellevue, Paris

LTI (1998) *Long-Term Integration of Renewable Energy Sources into the European Energy System*, Physica-Verlag, Heidelberg

Nakicenovic, N. and Messner, S. (1982) *Solar Energy Futures in a Western European Context*, WP-82-126a and WP-82-126b, International Institute for Applied Systems Analysis, Laxenburg, Austria

SolCat (2007) *Solar Catalonia: A Pathway to a 100% Renewable Energy System for Catalonia*, Fundació Terra, iSuSI, Ecoserveis, www.isusi.de/publications.html (English) or www.ecoserveis.net/ (Spanish/Catalan)

Chapter Five

Paths to a Fossil CO_2-free Munich

Stefan Lechtenböhmer

A city such as Munich can cut its CO_2 emissions by up to 90 per cent by mid-century by developing highly efficient building and mobility structures and adapted renewable and low carbon infrastructures. This is asserted by a recent study of the Wuppertal Institute for Climate Energy and Environment. Commissioned by Siemens it examined how a modern metropolis of 1.3 million inhabitants, such as Munich, third largest and one of the most dynamic German cities, can drastically reduce the amount of CO_2 it emits.

Munich 2058: Blueprint for cities leading the way to sustainable energy

Cities must lead the way to a climate friendly and sustainable energy future. Today, cities are home to 50 per cent of the world's population – a figure that is expected to climb to 60 per cent by 2025. The main drivers of their vibrant urban metabolisms, however, are fossil fuels such as oil, coal and natural gas. Cities directly and indirectly account for a majority of the global energy use and the respective greenhouse gas emissions.

But cities are not only the main driver of climate change. They will also bear the effects of global warming. In Munich, it is expected that among other effects the number of extremely hot days and tropical nights will significantly increase by the end of the century, imposing climatic conditions on the city that have not been known at least for several centuries here at the foot of the Alps.

Metropolitan areas represent both a high concentration of causes and consequences of climate change and a high capacity for action. High economic capacity, concentration of scientific and technological as well as economic know-how and decision-making competencies puts them into the pole position to develop the way to more climate-friendly and decarbonized lifestyles and economies. As one model of such a development, we have the city of Munich (Siemens, Wuppertal Institut, 2009; Lechtenböhmer et al, 2009). While it has survived almost 700 years, mainly based on regional renewable sources, the last 150 years like everywhere in the industrialized world have seen an increasing use of fossil energy. The resulting greenhouse gas emissions have risen to a current level of more than 8 million tons of CO_2 or 6.5 tons per capita. In our study we analysed two scenarios of how the city could again become almost fossil carbon free by its 900th anniversary in 2058.

Why cities should be almost carbon free by 2058

The *Fourth Assessment Report* by the IPCC (2007) has proved that climate change is the largest threat to human society and natural ecosystems, and preventing dangerous climate change is the foremost challenge for the world community. Its results also show that the risks of climate change increase with every degree of higher temperature increase over pre-industrial levels. Despite the international debate on adequate climate targets, it can be concluded that a reduction of global greenhouse gas emissions of 60 per cent or even more by the middle of the century will be necessary to prevent the most dramatic consequences of climate change (UN Foundation, 2007; Brundtland, 2007). This would mean that industrialized countries should reduce their greenhouse gas emissions by around 90 per cent and that global per capita greenhouse gas emissions should be well below 2 tons per capita by that time.

Table 5.1 Long-term greenhouse gas mitigation targets by selected cities

		City of Boston	City of Melbourne	City of Sydney[a]	London[b]	Munich[b]	New York City	Toronto	Zurich[c]
Base year		1990	1996	1990	1990	1990	2005	1990	–
Base year emissions	Mt CO₂e	7	3.5	2.3	45.1	10.2	58.3	22	–
Current emissions[d] in	Mt CO₂e	Nav.	3.8	3.6	44.3	9.8	58.3	23.4	1.4
tCO₂e/cap		Nav.	6.6	23.7	5.9	7.3	7.2	5.1	3.7
Target year		2050	2020	2050	2025	2030	2030	2050	2050
Reduction target		80%	100%	70%	60%	50%	30%	80%	70%
Target Emissions	Mt CO₂e	1.4	0.0	0.7	18.0	5.1	40.8	4.4	0.4
Baseline Emissions	Mt CO₂e	Nav.	4.5	Nav.	51	8.0	89.2	Nav.	Nav.

Note: [a] local government area; [b] Reduction target only for CO₂; [c] own calculations; [d] Data for 2000 to 2006. Note that base years, methodologies and scope of the targets differ between cities

Source: Lechtenböhmer et al.(2009); plaNYC A Greener, Greater New York; City of Boston Climate Action Plan; City of Sydney: Environmental Management Plan; City of Melbourne: Zero Net Emissions by 2020; Toronto: Climate Change Clean Air and Sustainable Energy Action Plan; Zurich: own calculations based on the Swiss Energy Research Concept

Many cities have already acknowledged the climate challenge and have adopted significant greenhouse gas emission reduction targets. Initiatives such as ICLEI, the C40 Initiative or the Covenant of Mayors have been formed to improve cooperation to foster municipal action (see www.iclei.org/; www.c40cities.org/; www.eumayors.eu/). However, so far only a small number of larger cities have set themselves far reaching targets, some of which could comply with the ambitious climate targets of 90 per cent greenhouse gas emissions reduction and more by the middle of the century, as can be seen from the table.

Scenarios for Munich

Munich's energy-related CO$_2$ emissions today amount to about 8.2 million tons, of which 46 per cent result from heating of residential, commercial and industrial buildings, a fifth of which is currently supplied by combined heat and power (CHP) via an extensive district heating grid. The rest is supplied mainly by single heating systems fired with natural gas and heating oil. Another 39 per cent of CO$_2$ emissions result from electricity of the municipal utility, which produces about 10 per cent more electricity than the city consumes in the local CHP plants, in a number of decentralized renewable units, in a share of a nuclear power plant and in some water power plants in nearby Alps. However, others also supply electricity to customers in the city. The rest of the CO$_2$ emissions taken into account here stem from passenger transport (12 per cent) and goods transport within the city borders (3 per cent). The study does not include the emissions from airborne transport nor freight transport outside the city borders or other emissions resulting from the import of goods and services to the city.

Given this scope, our study describes a comprehensive vision for an almost CO$_2$-free Munich. It sketches two scenarios for 2058, of which here only the more ambitious 'Target Scenario' is described. By using an all-embracing set of strategies and technologies in all fields it achieves a 90 per cent reduction of energy-related CO$_2$ emissions to about 750 kilograms (kg) per capita by 2050. This value probably meets the global aim to contain total greenhouse gas emissions per capita below 2 tons of CO$_2$ equivalent by 2050.

Seven key strategies for carbon freedom

The following key strategies have been proposed to reach the target of almost no fossil CO$_2$ by 2058:

- The main lever for fossil carbon free heat demand and supply, and also the largest lever in general, is the thermal improvement of residential and other buildings. By a rapid introduction of the currently most ambitious passive house standard for all new buildings and also for every renovation of existing buildings, energy use by this segment can be reduced by 80 per cent by 2058. To achieve this, however, virtually all buildings should be renovated over the next 50 years. This target is technically feasible but is a major economic and social challenge. Implementing less ambitious standards would mean that buildings with these standards will not become fossil carbon free by 2058 and thus delay the achievement of the targets to a point in time when they will be refurbished again. This fact has been reflected in a paradigmatic decision by the city council of Frankfurt/Main

(Germany). The city decided that all new buildings owned by the city or by housing companies controlled by the city (apart from some exemptions) have to comply to the passive house standard and – if possible – renovations should reach the same standard.

- The strategy, however, imposes an economic threat to the district heating (DH) system by massively reducing the amount of heat demand per building. Technologies for future central and decentralized DH systems have yet to be further developed. However, the study assumes that this problem will successfully be solved and CHP will be able to supply 60 per cent of heat demand and a third of the electricity demand in the Target Scenario.
- Additionally, renewable heat supply will be introduced by geothermal DH, biomass-based decentralized CHP plants and solar thermal appliances.

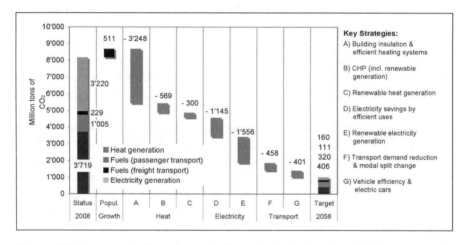

Figure 5.1 Key strategies for carbon emission reduction in Munich, 2058: Target Scenario

Source: based on Siemens, Wuppertal Institute (2009) and Lechtenböhmer et al (2009)

- The first strategy to achieve fossil carbon free electricity is electricity efficiency that could, by using mainly already existing highly efficient appliances, reduce per capita residential and office electricity consumption by almost 40 per cent.
- This will be flanked by additional renewable electricity, which will be partly produced locally, in the CHP plants, by water power and by photovoltaics, which could reach 400MW peak by 2058 by using about 40 per cent of the existing potential space for the installation of modules. About 40 per cent of the electricity in the Target Scenario will, however, be imported from offshore wind and solar thermal power plants – a strategy that has been already introduced by the municipal utility. About 40 per cent of the wind capacity envisaged for 2058 will be on the grid in the German North Sea as

early as 2013, and a decision has been made to buy a share of a projected solar thermal power plant in Southern Spain.

- The core strategy of the Target Scenario towards fossil carbon free transport is to gradually change mobility patterns to slightly reduce average transport distances and increase the share of public transport (within the city, but also important on long-distance trips) (assuming public transport is not coal-fired electricity), walking and biking.

- More efficient vehicles and a shift in urban transport towards electric vehicles contribute to significant energy savings and indirectly to an increased share of biofuels and renewable electricity in the transport fuel mix. The Target Scenario assumes that about 80 per cent of motorized inner-city individual mobility will be covered by electric vehicles, which will be either specific small vehicles for short distances or plug-in hybrids. These will run on shorter distances with electricity and outside the city with fuels.

The role of renewable energies

The scenario for Munich shows that a comprehensive package of measures and technologies, which is mainly already available on the market or at least 'in the pipeline' is available to bring cities on track to become virtually fossil carbon free over the coming century and to cope with the targets set to slow down climate change and to prevent its worst damages. As the seven key strategies show, the first priority of the scenarios is to drastically increase energy efficiency, i.e. make the most of the energy used by the urban metabolism by highly efficient buildings, adapted supply systems and cascading (first for electricity then for heat, or first for process heat then for DH) uses of energy. By strict implementation of these principles, primary energy demand of the city of Munich can be reduced by more than 70 per cent, despite a slightly growing population. The remaining energy demand can be supplied by more than 60 per cent from a mix of local and regional as well as imported renewable energies, more than half of it coming from renewable electricity and the rest from renewable heat and biofuels. And, given the fact that the total amount of renewable energy supply will reach only 5200kWh or about 450 litres of oil equivalent per capita per year, we assume that this amount can be supplied in a sustainable manner.

Conclusions

The question of whether human society will be able to cope with the challenge of manmade climate change by developing fossil carbon free infrastructures and economies will be answered in the cities. They are not only core to the problem but also to the solution. For the southern German metropolis of Munich, our study demonstrates that energy efficient and low carbon technologies and infrastructures are already available and could be rolled out for general use.

However, reshaping urban infrastructures comes at a high cost. Converting all buildings in Munich to low energy passive buildings would need €13 billion in additional investment at the current price level, or €200 per capita per year. It is, however, an investment that will *most probably* prove to save more by reducing energy imports than it costs.

This means investing now in the right strategy that will offer multiple chances for the cities:

* Jobs and markets can be created through the investment in fossil carbon free infrastructures.
* This investment is an effective hedge against future increasing prices and fossil fuel supply risks.
* Defining a strategic target to develop a nearly fossil carbon free city can provide further first-mover advantages, ranging from the creation of a positive image for the city to creating business opportunities in the emerging growth market of sustainable urban (infra)structures.

The study shows that transforming a city into a virtually carbon-free urban environment will be a significant challenge. To master it the aim of virtually carbon-free cities has to become top priority for all stakeholders: decision-makers, utilities, urban planners and, particularly, investors and residents.

Acknowledgements

I would like to thank Dieter Seifried, Claus Barthel, Susanne Böhler, Rüdiger Hofmann, Kora Kristof, Frank Merten, Frederic Rudolph, Clemens Schneider and Dietmar Schüwer who are the co-authors of the study briefly reported here. I also thank Stefan Denig and Daniel Müller from Siemens and their colleagues who strongly supported the research for this study and contributed valuable inputs, as well as Tim Schröder.

References

Brundtland, G. H. (2007) 'Speech at the Gleneagles Dialogue', 3rd Ministerial Conference, Berlin, 10 September 2007, Bundesministerium für Umwelt, Naturschutz und Reaktorsicherheit (BMU)

IPCC (2007) 'Summary for policymakers', in B. Metz, O. R. Davidson, P. R. Bosch, R. Dave, and L. A. Meyer (eds) *Climate Change 2007: Mitigation. Contribution of Working Group III to the Fourth Assessment Report of the Intergovernmental Panel on Climate Change*, Cambridge University Press, Cambridge and New York, NY

Lechtenböhmer, S., Seifried, D., Barthel, C., Böhler, S., Hofmann, R., Kristof, K., Merten, F., Rudolph, F., Schneider, C. and Schüwer, D. (2009) *München 2058 – Wege in eine CO₂-freie Zukunft, Hintergrundbericht*, Wuppertal, forthcoming

Siemens AG, Wuppertal Institute (2009) *Sustainable Urban Infrastructure, Munich - Roads Toward a Carbon-free Future*, Munich (in press)

UN Foundation, Club de Madrid (2007) *Framework for a Post-2012 Agreement on Climate Change. A Proposal of Global Leadership for Climate Action*, 10 September, Club de Madrid

Chapter Six

100% Renewable Champions: International Case Studies

Anis Radzi

This chapter chronicles the trials and successes of 14 renewable energy projects around the world, with a special focus on islands, regions and cities, in their quest for 100 per cent self-sufficiency based exclusively on renewable energy sources. While some are works-in-progress, others have achieved over and beyond their original targets. Each project tells a story, a narrative illustrating the human traits of inspiration, dedication, drive and initiative, as the overriding means to achieve an ecological society and economy free from fossil fuels. Collectively, the projects illustrate processes entailing more than mere technics or money.

El Hierro, Canary Islands

Figure 6.1 Juniper Tree at El Sabinar, El Hierro, Canary Islands

Source: Eckhard Pecher

Basic facts

El Hierro is the smallest island of the Canary Islands, with a surface area of 276km² and a population of 10,600 people. More than 60 per cent of the territory is classed as nature reserve (Iris Europe, 2007). By 2010, the island hopes to derive all of its energy from wind, water, solar and silage power. Hence, becoming the first European island to cover all of its energy needs with clean and renewable energies by opting for a development model based on conserving the environment.

The first phase of El Hierro's plan will involve a 9.9MW wind-hydro scheme, whereby desalinated water will be pumped by windpower generated by a 7.35MW wind farm to a reservoir created at 700 metres above sea level – this will result in the flooding of the crater of an extinct volcano. When required, water will be released to the lower reservoir driving turbines to generate hydroelectricity. Besides filling the reservoir, the water desalination plant will help compensate for any evaporation losses and produce water for irrigation and domestic use (INSULA, 2008). The island's existing 8.3MW diesel power plant will remain in use as a contingency source of energy. In addition to the hydro-wind station, El Hierro has plans for more solar thermal installations, solar photovoltaics, several biogas/biomass plants and an integrated alternative transport system.

How it evolved and what drove it

In the early 1980s, the Cabildo de El Hierro (island government) needed a development model that respected the island's heritage, conserved the natural resources, improved basic infrastructures, advanced communications in and out of the island and fostered cooperation among industries. It wanted to move away from mass tourism based on real estate. But more importantly, it wanted to prove that a sustainable development model on the island was possible (Padrón, 2004).

When the island was declared as a World Biosphere Reserve by the United Nations Educational, Scientific and Cultural Organization (UNESCO) in 2000, the El Hierro government took advantage of the momentum gained from this to push through its new Island Planning Regulations (PIOH) as well as launch its sustainable development plan, entitled the 'El Hierro 100 per cent Renewable Energies' project. A first for the Canary Islands, the plan aimed at making El Hierro the first in Europe to be supplied with renewable energies, turning the island into a worldwide benchmark for implementing energy self-sufficiency and autonomy systems based on clean energy sources on isolated islands.

The first public consultation meeting was held in April 2004 with various socio-economic groups on El Hierro to help explain the importance of the project to the development of the island and to highlight the significance of community involvement in guaranteeing its success. Locals were invited to express any concerns or ideas that they may have regarding the development.

They were informed that training sessions would be made available so each individual would be able to easily adapt to the new technologies and organizational structures as well as be prepared for the responsibility of fixing and maintaining systems.

The turn towards wind-hydro power was natural for El Hierro because of the strong trade winds, rugged terrain and low electricity demand in comparison with the other islands of the Canary Archipelago. It is estimated that wind power alone could potentially generate more than 80 per cent of the island's own electricity.

With financial support from the Directorate-General for Energy and Transport (DG TREN) of the European Commission, a consortium of seven partners called Gorona del Viento El Hierro, which included the island government, the Canary Islands through the Technical Institute of the Canary Islands (ITC) and the local utility (UNELCO-ENDESA) was established. By 2007, a call for tenders was published for the construction of the wind-hydro plant. Its financing will be greatly assisted once island residents become directly involved in the project as co-owners of the station by partly purchasing shares in the company. Once in use, studies into the day-to-day running of the facility will be conducted to determine the feasibility and economy of such systems for their replication on other islands (INSULA, 2008). The plant is scheduled to be fully operational by late 2010.

As part of the island's energy plan, a new transport programme will synergize a range of alternative transport systems through the support of a local transport cooperative. An integrated network will include a hybrid bus as an addition to the local fleet; an electric, battery-powered minibus for mixed tourist-public use, which is recharged at a photovoltaic station; a revised pedestrian network and an information transport system called the 'El Hierro – Digital Island'. A new ticketing system will turn private vehicles into public transport by using an electronic system for fare payments, which will help save energy and resources in the transport sector, particularly in dispersed rural areas (IDAE, 2001a).

Already in existence is El Hierro's solar thermal energy programme or PROCASOL, which was created to promote solar thermal energy installations on the island. The programme has helped launch a local company whose task was to substitute electrical heaters with solar thermal systems in order to reduce the total electricity demand for domestic hot water. The programme, managed by ITC combine direct funding and loans at a zero interest rate in three years with monthly instalments, both granted by the Regional Ministry for Presidency and Technological Innovation and ITC, as finance. The direct subsidy is €120 per m² installed collector area. The owner of the system only has to pay an initial amount of around 10 per cent of the total installation in advance; the remainder is financed by ITC at zero interest (Piernavieja et al, 2003). Other incentives include guarantees for the collectors in their installation and maintenance. The ultimate goal of the programme is the installation of 2500m² of collectors in order to cover the entire island market.

Previously, photovoltaic energy came only from a few small stand-alone systems (7kWp), disconnected to the grid. The main barriers to previous implementation were insufficient information, the lack of qualified suppliers/installers on the island, the shortage of private initiatives and the apathy of locals. Some of these barriers were lifted by awareness campaigns, training courses, workshops and seminars. But to further increase the percentage of renewable energy into the grid, ITC created a local company to supply and install PV systems, launching the so-called '10 PV roofs', which involved the installation of ten 5kWp units on public buildings (INSULA, 2008).

As part of the 'El Hierro – zero waste' initiative, the island produces biogas from sewage sludge, animal waste, organic municipal wastes and organic industrial waste from a slaughterhouse and dairy. This biogas programme is the result of an international initiative with the island of Cuba, which has been relying on technical experience and training gained from El Hierro. The first phase of this project has resulted in the installation of several digesters sponsored by El Hierro's island government (IDAE, 2001a). Indeed, biofuels have already partially substituted the diesel used in the power station or in some transports on the island. The waste from forests and plantations (banana, pineapple) has also been used as fuel for heat production, reducing the use of electricity and butane. Future plans will involve the recycling of used oils and energy crop cultivation as alternative fuel sources.

Informing the island residents of the government's energy plans involved leaflets, brochures, publications, website and software development, organization of workshops and seminars and attendance in conferences and exhibitions. Technical visits were organized composing of local authorities, consumers associations, industry and private investors, tourism industry, interested citizens from islands worldwide and EU representatives among many others (INSULA, 2008).

Lessons learnt

El Hierro faces one major environmental challenge: the wind-hydro power station will result in the loss of La Caldereta and the volcanic cone of Las Tijeretas, two topographical and geological attractions to be transformed into reservoirs. The island government is conscious of this issue and is working to integrate future installations into the landscape with more care in order to minimize visual and environmental impacts (Piernavieja et al, 2003).

A technical challenge for island will be the fluctuating characteristic of wind, which the government hopes to counter by storing energy and by setting a limit of 12 per cent market share for wind energy in the Canary Island electricity act. The island recognizes that in order to fully realize the potential of wind power, the energy must be stored. It will achieve this by taking advantage of the new wind-hydro station. The energy from excess wind power is used to pump water into a reservoir several hundred metres above sea level;

whereby water is released to lower turbines to be converted to electricity when the wind power is in short supply. A stable frequency and voltage in the grid is thus achieved from the energy storage and the manageable power output from the hydro turbines ensures supplies whenever required (INSULA, 2008).

According to the Canary Island government, the hydro-wind power station will save around 6000 tons of diesel fuel per year, oil that will no longer have to be imported by ship and represents a saving of over €1.8 million a year on the diesel oil bill. This means a reduction each year of CO_2 emissions by 19,000 tonnes, sulphur dioxide (SO_2) emissions by 100 tons, nitrogen oxide (NO_x) by 400 tons and particulates by 7 tons (Padrón, 2004).

Through the dedication of the local government and various organizations such as UNESCO, the island of El Hierro will be able to replace its fossil fuel use with renewable energy sources and prove that renewable energy integration can provide 100 per cent energy supply on islands, even into weak grids within isolated areas, with pumped water storage as an economic way of accumulating energy. It will set an important precedent for all islands worldwide as they benefit from the experience gained in El Hierro. The inhabitants of islands will enjoy many environmental and economic benefits in form of reduced CO_2 emissions, a better quality of life, energy independence and increased employment.

The Pacific Islands

Figure 6.2 An improved ram pump technology brings running water to hillside villages in the Philippines. Ram pump designed by AID Foundation and installed with the help of villagers in Negros

Source: Ashden Awards/Martin Wright

Basic facts

Solar photovoltaics have been used for more than 20 years in the Pacific Islands. The total off-grid capacity installed in the region today is around 350MW. In Kiribati, for example, solar panels represent 20 per cent coverage on the island and are used to power rural health centres and remote radio-telephone sites (Mitra, 2006). Indeed, renewable energy techniques differ from island to island. In Fiji, a village cooperative runs a small-scale hydroelectric project to provide electricity to over 200 homes. In Papua New Guinea, a hybrid renewable energy system, involving wind turbines and solar panels, provides an independent power supply to a school. In Samoa, a medium-sized hydro-project in Afalilo on the capital island of Upolu feeds 80 per cent hydro and 20 per cent diesel to a hydroelectric power plant. In the Cook Islands, coconut oil is used as fuel in a normal diesel engine, and has been used in Vanuatu to fuel buses, taxis and other vehicles, as well as generators that power a hydroponics project (UNESCO, 2003).

How it evolved and what drove it

As early as 1985, conventional electrification was difficult, expensive and generated very little demand. To deliver and reticulate diesel fuel was costly and unreliable and there were often problems with engine and grid maintenance on remote islands. So the use of solar photovoltaics became the most practical and cost-effective alternative for rural electrification (Marconnet, 2007).

The European Union (EU), France and the US were among the first to offer assistance in the implementation of solar electrification, which was much welcomed by the island nations. There was no model to follow so the Pacific became a laboratory for renewables-based rural electrification, with trials of various technologies such as biomass gasification, biomass combustion, biofuels, biogas, wind, tidal and solar power (United Nations, 2004).

Most of the early projects were small, with the largest including only a few hundred users. Local training programmes were provided by various organizations and supplemented by regional training activities. By 2000, plans were in place for large-scale development of renewable energy for rural electrification but also to replace conventional energy for grid based power generation and transport (United Nations, 2004).

A range of funding agencies provided substantial technical and financial support to renewable energy and efficiency projects in the Pacific region. Australia is the largest aid donor to the Pacific Islands. In 2000, in association with France, Australia set up a Pacific Renewable Energy Programme to assist the island nations to improve energy efficiency measures, increase the use of renewable energy and reduce emissions through fuel switching. As the eighth largest supplier of aid, the EU earmarked €15.2 billion (about $14,000 million) to be distributed over twenty years. Their financing moves towards budgetary aid and away from projects. In Tuvalu, for example, they used the government's existing arrangements for the finance they had available for the activities they wanted to support rather than engaging contractors themselves (ICEPAC, 2000).

However, by 2003 it was estimated that 70 per cent of people living in the Pacific Islands still did not have access to electricity. Despite the abundance of renewable energy resources such as solar energy, hydro, bio and wind, the islands remained dependent on imported diesel for their power and transport needs, the cost of which was almost four times that of neighbouring countries such as Australia and New Zealand (ADB, 2003). With rising petroleum prices as well as rising sea levels over low-lying atolls, the situation was becoming increasingly unsustainable (Woodruff, 2007).

In 2007, the World Bank and the International Finance Corporation (IFC) decided to implement the Sustainable Energy Financing Project. Their donation of $9.5 million aimed to fund solar photovoltaics and pico-hydro projects over ten years. Primarily backing the private sector, the World Bank began by making available loan guarantees and loans to people in rural areas to buy solar equipment.

As a result of all this effort, the inhabitants of the Pacific Islands today enjoy reliable and affordable electricity through better quality lighting and appliance use. Shopkeepers can operate at night as street lighting has improved. There is an increase in productivity as household members can engage in productive activities such as weaving or studying for longer hours at night; hence improving income and education. Hygiene has improved through better food storage and reduced indoor pollution levels. There are more entertainment and communication opportunities through the use of televisions, mobile phones and radios. The reduced risk from fires associated with the use of kerosene lamps and fuel wood has ensured personal safety. A study in 2006 showed that solar home systems and micro-hydroelectricity were gradually replacing the use of diesel, kerosene or dry cell batteries as the renewable energy systems were becoming more and more cost-effective (Woodruff, 2007).

Lessons learnt

The development of renewable energy in the Pacific Islands faced a variety of institutional, financial, educational and technical challenges. First, the renewable energy sector in the region often received low political priority. With the exception of the island of Palau, most countries failed to develop policies that ensured that renewable energy technologies were adequately considered during energy planning. When electrifying rural areas, for example, it was difficult to manage at the village level as well as regulate the involvement of private companies. Some countries did not even have a policy for rural development. It was also difficult to implement energy efficiency programmes because of the lack of building codes, particularly in the larger island countries (UNESCAP, 2002). The success of any renewable energy projects required management procedures that made certain the maintenance of equipment and the collection of consumer fees that were set to an appropriate level which guaranteed the financial sustainability of projects (Woodruff, 2007).

With regards to market barriers in the Pacific Islands, there were incentive problems, high transaction costs and 'lean-and-incomplete' electricity markets. Government policies were needed to lower the high up-front costs. Some have suggested such strategies as import tax exemptions, the provision of soft loans, or enhancing the consumers' ability to pay for energy through electrification projects that incorporate income-generating schemes (Woodruff, 2007). In comparison with policy tenders and renewable trading certificates, tariffs proved to be most successful for connecting small- and medium-scale renewable energy independent producers to the electrical grids (Marconnet, 2007).

The generous financial donations for the renewable energy projects in the Pacific Islands were often distributed in a fragmented and discontinuous way, due to ad hoc coordination and unclear responsibilities between energy agencies and governments. The failure of many projects was often the result of unrealistic goals that were either too ambitious or vague. Activities were often incompatible with existing policies and socio-economic structures. Furthermore, as projects were largely grant-funded, this created a low sense of ownership among locals, hence the issue of over-reliance on external aid to fund projects (ADB, 2003).

The need for public education on renewable energy was critical, particularly in the outer islands and rural areas, where a number of solar electrification projects have failed due to insufficient information and hence poor maintenance (Marconnet, 2007). And despite the involvement of many organizations in providing renewable energy training, the technical expertise to plan, manage and maintain renewable energy and energy efficiency programmes were still inadequate. Projects suffered from a rapid turnover of field maintenance personnel, resulting in few qualified trainers. Individuals who received very good training frequently left for better-paid employment elsewhere.

The fragile marine ecosystems of the Pacific Islands will be effected by any significant infrastructural change (Woodruff, 2007). Wind turbine installations, for example, involve deep localized excavation, the use of a substantial amount of concrete for its base construction and the transportation of heavy components. The turbines often produce mechanical and aerodynamic noise and cause electromagnetic interference in televisions and radios as well as interference with bird migration and low flying aircraft. While owners see wind power as a sign of prosperity, neighbours consider them a disturbance.

The Pacific experiences reasonably good winds, but these are generally non-existent in equatorial areas. For this reason, as well as the lack of available land, the difficulty in obtaining land leases, and the large risk of damage to turbines posed by cyclones, large-scale wind farms have not been feasible

(Woodruff, 2007). Wind turbines only provide power to the electrical grids in Fiji, Nabouwalu, Nabua, Butoni (10MW) and the island of Mangaia in the Cook Islands.

Most of the early solar PV systems in the Pacific suffered from technical deficiencies, not in the panels themselves, but in controllers, batteries and other associated appliances. Among other causes were unreliable components, inappropriate design, such as undersized panels, improper installation and poor maintenance (Liebenthal et al, 1994). While most flaws have been remedied, one minor challenge remained: light bulb failure – the replacement of which required several days of waiting for households. Nonetheless, in isolated areas where skilled technicians were scarce, photovoltaics systems still worked best. They required less maintenance compared with diesel generators, which needed regular and costly overhauls, sometimes taking weeks to months to repair when it fails.

Hydroelectricity has suited larger island nations due to the high start-up costs and high electricity thresholds the facilities demand. Smaller-scale micro-hydro schemes adapted better in areas where populations were low and isolated, and where geographies could not support large water catchments. Although small, micro-hydro was not without its challenges. The use of local wood for dam construction caused the operational failure of one hydro plant in a small rural village, left dishevelled because all efforts were focused on maintaining the plant. Despite the use of local resources, the local timber in this case was not sufficiently durable.

The use of coconut oil as a biofuel has been limited by operational and market factors. Because of the viscosity of coconut oil, incomplete combustion causes carbon deposits in diesel engines – a problem that can be avoided in indirect injection engines (Vaitilingom, 2006 cited in Woodruff, 2007). The production of coconut oil itself has suffered not only from fluctuating prices, weak management and limited investment but also from the limited availability of copra, rising labour costs and natural disasters (CocoGen, 2005 cited in Woodruff, 2007). The feasibility of coconut oil as a biofuel requires a resurgence in the copra industry alongside improved processing technology.

In summary, access to energy has improved the health, education and economic development of the Pacific Island nations, helping to ease environmental problems already in progress. However, the degrees of success will depend on local conditions, a good mix of renewable resources, availability, credible management and stable financial frameworks.

Samsø, Denmark

Basic facts

The Danish island of Samsø is located off the coast of the Jutland Peninsula. Its population density is low: some 4100 people live on 11,400 hectares. Agriculture is the biggest sector of its economy and tourism the second largest, attracting half a million guests a year. At the EU Renewable Energy conference

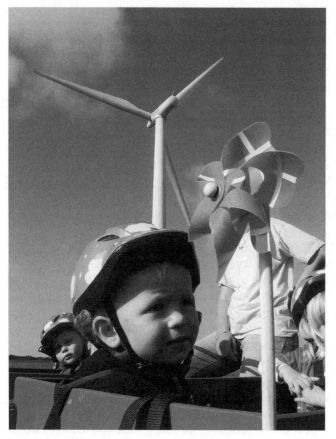

Figure 6.3 Daycare by bike in Samsø

Source: Samsø Energy Academy

in Toulouse in 2000, Samsø received an award for being 'the best renewable energy island in Europe' (INSULA, 2001). Today Samsø is known as the inspiration behind other 'renewable energy islands' such as El Hierro in the Canaries, La Maddalena in Italy and Aran Islands in Ireland, each focusing on different kinds of renewable energy.

How it evolved and what drove it

In 1997, the Danish Ministry of Environment and Energy sponsored a renewable energy contest challenging communities to abandon fossil fuel dependency in a planned fashion, as part of the Danish Action Plan of 1996. Chosen from five competing islands, the winning plan proposed that the island of Samsø power and fuel itself entirely on renewable energy within a ten year period. Much faith was placed in a plan that was devised by an engineer who did not actually live on the island, albeit some consultation with the island's mayor. Limited funding caused the project to go nowhere for many years, as Samsø received nothing for winning, until a federally funded position opened to head up the initiative. Søren Hermansen, an environmental studies teacher, secured the position.

Reaching out to the community took considerable time and effort. Hermansen made a point of attending every local meeting to pitch the project. It wasn't until he proposed the project as a social relations undertaking – asking community members to think about what it would be like to work on it together and have something to be proud of – that the initiative began to appeal to the citizens of Samsø. With increasing success and popularity, seminars on wind energy were held and soon energy development cooperatives began to emerge (Kolbert, 2008).

The Renewable Energy Island plan for Samsø was implemented with five goals. They were to save energy and increase efficiency; expand the collective heating supply systems fuelled with renewable energy; expand individual heating systems using heat pumps, solar heating or biomass; establish land based and offshore wind power plants; and gradually convert the transport sector from petrol and oil to renewables-based electricity (INSULA, 2001).

The process of finding wind turbine sites and financing began in 1998. The Samsø Energy Supply Company was established to coordinate the various energy supply projects. Eleven 1MW land-based turbines were erected between 1999 and 2000 at a cost of DKK66 million (about €8.8 million), with shares sold to islanders, generating local income with local power. The number of turbines had to be reviewed in a public hearing after 40 private applications for wind permits were submitted (INSULA, 2001).

Ten more generators, rated at 2.3MW were installed offshore in 2002 after five years of detailed planning at a cost of DKK250 million (about €33.3 million), funded by the Danish Energy Agency (Samsø Energiakademi, 2008). They were installed as a notional, imperfect compensation for the continued use of fossil fuels used in cars, trucks and ferries. The turbines would also supply electric cars and hydrogen fuel cell cars in the future. The new turbines soon helped Samsø achieve 100 per cent of its electricity demand, compared to 75 per cent as previously planned (INSULA, 2001).

Most windmills were funded by individual investors, such as local farmer Jøgen Tranberg; and only two were owned collectively, in contrast to common perceptions about Danish traditions of wind power collectives-based investment. Annual donations received from the windmill owners contributed to other public energy projects on the island.

By 2001, fossil fuel consumption had been cut in half and by 2005 Samsø produced more renewable electricity than it used so it began exporting it. Since then, heat has been generated in the DH plants, cleanly combusting locally grown straw and wood chips, producing superheated water in four central plants, to be pumped underground to homes. The plan was that 60 per cent of homes on the island would be heated through DH systems, and 40 per cent of individual homes. Over 2000 homes, those outside the DH areas, were given various heating and power improvement options, from conservation to renewable energy installation (Samsø Energiakademi, 2008).

Samsø's pioneering Tranebjerg district has produced straw-based heat since 1994. By 2002, the Nordy/Mårup DH station was opened to produce 80 per cent of the heat from wood chips and 20 per cent from a 2500m² solar heating system. Homeowners with oil furnaces were the primary target and they were asked to voluntarily sign up by June 1999 and only pay a membership fee of DKK100 ($20). To provide a disincentive to postponing change, the cost after that date rose to nearly $8000. As a result, 78 per cent of homeowners in Nordy signed up and 83 per cent of homeowners in Mårup. Only the new buildings built in areas with existing or planned DH were obliged to connect to the DH system. In 2003, the straw-based Onsbjerg plant was opened by a local contractor and run by a local committee, and in 2004 the Ballen-Brundby straw-based plant was opened as a heating system owned exclusively by the consumers themselves (Samsø Energiakademi, 2008). To date, the four plants cover 70 per cent of the islands heating needs (INSULA, 2001).

To reduce the heating requirement on the island, buildings were renovated or retrofitted with new insulation. Companies and public buildings were fitted with new energy control and monitoring systems. Pensioners were eligible for refunds of up to half their energy conservation costs from the national government, with a maximum of DKK25,000 ($4705). This resulted in new insulation work and window installations worth nearly DKK3 million ($565,000), after 92 island pensioners agreed to participate in the programme (INSULA, 2001).

The total investment for the renewable energy island projects was about DKK425 million or €57 million (Samsø Energiakademi, 2008). Financial resources primarily came from the Danish Energy Agency, Aarhus Regional Authority and Samsø municipality. Other contributers included the Samsø Business Forum, Samsø Farmer's Association, Samsø Energy Supply Company and Finland's Ministry of Energy. It is important to note that in the earlier stages the government offered no funding, tax breaks or technical expertise (Walsh, 2008).

Lessons learnt

There were other types of renewable energy installations planned on Samsø but most had failed. Among them were plans to use excess heat from the Sealand ferry for DH purposes, to extract biogas from pig farms' slurry and to produce natural gas from cow manure. The one project that did succeed was a biogas facility based on methane extraction off a disused landfill site. Methane gas is used to run a 15kW motor/generator, which produces electricity that is then sold to the grid. Installed in 2000 by a local cooperative that included the Samsø Energy Company and some farmers, it received financial support from the Danish Energy Agency. The success of this installation has resulted in a similar facility at Samsø's existing disposal site, which depending on the volume and quality of the methane gas, generates not just electricity but heat for site buildings (INSULA, 2001).

The transport sector has been difficult and costly to regulate because much of the sector relies on oil. The initial energy plan from 1997 recommended more energy efficient driving habits, more flexible bus transportation service and the use of electric cars, but in the case of electric cars, the market has been nonexistent (Samsø Energiakademi, 2008). The municipality's experiment with electric cars failed when one of the demonstration vehicles spent most of the year with the mechanics (Kolbert, 2008). To begin to abandon fossil transport fuel, many farmers today experiment with biomass and have begun to run their tractors on rape seed oil. But some have required expert help in relation to pressing oil, expertise that has not been readily available on the island.

Within an eight-year period since 1997, heat production from renewable energy in Samsø increased from 25 per cent to 65 per cent and heat consumption decreased by 10 per cent. Generally, the share of renewable energy increased when oil imports dropped. Unfortunately, electricity consumption has remained the same because despite savings and better practices in energy use, homes have more domestic equipment (Samsø Energiakademi, 2008). Households that had better insulation also tended to heat more rooms, so the net savings were zero (Kolbert, 2008).

The promotion of renewable energy by public campaigns and local effort has helped increase the rate of renewable energy installations on Samsø. About 300 private homes have invested in individual renewable energy heating systems. Thermal solar units have been installed on roofs for hot water and space heating, while oil furnaces have been replaced with biomass boilers or heat pumps, the latter as either large ground-heat systems to supply the entire house or smaller air-to-air heat pumps for localized space heating (Samsø Energiakademi, 2007). There are also solar installations in ports, at a youth hostel, campsite and a holiday camp (INSULA, 2001).

The renewable energy projects (windmills, DH plants, renewable energy installations in private buildings) have been an important source of local jobs. In 2007, the Samsø Energy Academy was opened to house the Samsø Energy Agency, Samsø Energy and Environment Office and Energy Service Denmark. It showcases Samsø's renewable island project and works as an educational and research laboratory (Samsø Energiakademi, 2008).

Samsø's ecotourism has boomed, with many people coming to visit the Renewable Energy Island project. They range from politicians, businessmen, students, planners, engineers, schools and even some activists. The Welcome Center was opened by the Samsø Ecomuseum in 2000, enabling tourists to explore the cultural history of the island as well as learn more about the renewable energy island project.

The success of Samsø's renewable energy island was the adoption of the project, empowerment and ownership by the people. One dedicated local citizen was able to reach out to his community and inspire it to enact local but powerful change. Citizens came together to create local organization and to adapt to new lifestyles based on renewable energy. With a diverse mix of renewable energy sources combined with energy efficiency and carefully structured sustainable development measures, the island has been able to demonstrate that self-sufficiency is possible for entire communities.

Güssing, Austria

Figure 6.4 The 'Green Drop' symbolizing Güssing's ökoEnergieland model

Source: EEE GmbH

Basic facts

Güssing is located in a densely-forested region of south-eastern Austria, along the Hungarian border, 130km southeast of Vienna. With a population of 4000 people, the town features a 12th-century castle built by Hungarian nobles. Traditionally Güssing's economy relies on small-scale agriculture, with farmers selling corn, sunflower oil and timber. Until the fall of the Iron Curtain, Güssing was one of the poorest areas of Austria. It had neither railroad nor highway, and poor infrastructure resulted in high unemployment with many residents migrating or commuting far to work elsewhere. It was not until 1974 when it was given town status (ökoEnergieland, 2008).

In 2004, Güssing received the European Solar Prize in recognition for its sustainable regional development process (Eurosolar, 2004). In September 2005, its European Centre for Renewable Energies (EEE) was awarded the

Global 100 ECO-TECH Award in Japan for the 'Güssing model', winning ¥1 million in prize money. Chosen from more than 230 entries worldwide, the award recognizes projects that have made a significant contribution to solving environmental problems and to creating a sustainable future.

Today Güssing is a model for environment-friendly energy production based on the principles of energy saving, value-adding and environmental protection (Rauch, 2005). It has become the first town in the EU to cut carbon emissions by at least 90 per cent.

Güssing's achievements can be attributed to the 30 renewable energy plants operating within 10km of the town. These plants cover the town's three main energy demands: fuel for transportation, residential heating and electricity (Aichernig et al, 2002). The €2 million rape oil refinery produces 8 million litres of biodiesel per year by processing rapeseed and used cooking oil. The biomass plant supplies 23MW of heating to the DH network by processing 24,000 tons of agricultural and forestry material. A biogas plant representing an investment of €2.1 million generates 0.5MW of electricity and 0.6MW of heat by using 11,000 tons of silage off 250 hectares of farmland. The grass and maize is anaerobically digested in the biogas plant to generate gas, which is fed into the plant to make synthetic natural gas. This is compressed and sold to vehicles on the main arterial road.

Güssing has two solar demonstation units consisting of a photovoltaic plant that produces 9MW of electricity and a solar thermal plant that generates 15MW of heating per year. The plants use innovative solar roof tiles made from recycled plastic waste, which are integrated with solar modules. However, the most important innovation in Güssing has been the state-of-the-art biomass gasification plant. With a fuel capacity of 8MW, the plant operates 8000 hours a year to produce 2.5MW of electricity and 4.5MW of heat for DH by processing 2300kg of wood per hour (RENET, 2006). The heating station also operates drying chambers, with scope for district cooling.

Güssing's consumers consist of 300 private houses, public offices, schools, hospitals and industrial companies. The plants generate a total of 22MWh per year. An excess of 8MWh is fed into the national grid, generating €4.7 million in revenue each year and a profit of €500,000, which is reinvested into the community and its renewable energy projects (GreenUpAndGo, 2007). Compared to the €6.2 million spent on energy in 1991, the municipality generates around €13 million each year from the renewable sector (based on 2005 figures), with 1000 new jobs created and 60 companies attracted to the area.

How it evolved and what drove it

The renewable energy initiative began in Güssing primarily due to economic need. By the late 1980s, the municipality had a massive fuel debt. The town was struggling to pay its €6 million electricity bill. Fossil fuels bought from the outside were very costly (GreenUpAndGo, 2007). The debt had to be cleared before any form of investment into renewables. But it was only after the sight of the forests, farmland and sunshine from the top of Güssing castle which finally convinced the town's mayor, Peter Vadasz, of the town's renewable potential (Alden, 2007).

With the help of Rheinhard Koch, the town's electrical engineer, and Herbert Sattler, the head of the local timber growers association, they began examining ways the residents could benefit from their natural resources. In 1990, a model was developed to completely phase out energy supplies that were based on fossil fuels. The plan was to supply the town of Güssing with renewable energy and then the entire region. First, energy saving measures were introduced, including better windows and full thermal insulation. Then, public buildings were ordered not to use any more fossil fuels. This resulted in a reduction of 50 per cent in energy expenditure within only two years.

The construction of the bioenergy plants soon followed. The biodiesel plant was put into production in 1991, producing more biodiesel than needed, and within the following four years, the biomass and biogas plants had been installed powering and heating local homes and buildings in the centre of town. When the biomass gasification plant was installed in 2001, it enabled Güssing to become truly energy self-sufficient (BMVIT, 2007).

Work on the DH system began by digging up the town's streets to take the insulated pipes carrying the hot water (which was heated in the central boiler of the biomass plant) to the whole town, street by street, even to homes that did not want the supply, so they could join later whenever they wished (Douthwaite, 2006). Heat is recovered from the hot water through heat exchangers in the buildings of customers. The cooled-off water is returned through separate pipelines (Tirone, 2007). The benefit of this system was that households no longer needed individual boilers. The new system would be just as convenient as heating with oil and would not be costly, although some residents remained unconvinced. Today the DH system is 30 per cent cheaper and more efficient. A 5MW boiler was installed in 1998 to cope with the growing demand, with several extensions since resulting in a 42MW capacity system along a 27km network.

In the next project, it was decided that in addition to heating, Güssing would produce electricity and synthetic fuels. In 1998, the Viennese scientist, Hermann Hofbauer and Vienna's Technical University were asked to build a demonstration plant applying an innovative technology, which were unlike the typical gasification plants around the world. Developed by Hofbauer, this type of plant avoids the build-up of tar when wood is heated. The process involves producing natural gas by using steam to separate carbon and hydrogen from scrap lumber, and then recombining the molecules. The natural gas fuels an engine that produces electricity and the by-product heat is used to produce warm water for the DH system. The gas is also used in the production of synthetic diesel oil, through the Fischer-Tropsch process (Tirone, 2007).

The main advantage of the gasification process is that it uses steam rather than air to turn the biomass into a gas with very little tar and with a high heat value. There are also low gaseous emissions and no liquid emissions (Aichernig et al, 2002). The equilibrium achieved between the combustion and gasification reactions, which occur automatically, ensures stable operation without any excessive regulation and adjustment (RENET, 2006). Another crucial advantage is the facility's compact construction.

In economic terms, the biomass gasification power plant has been a highly successful model. Electricity sold to the electrical grid operator is €0.25 less than the €0.373/kWh previously paid by domestic consumers in the area (Douthwaite, 2006). The biomass feedstock is costly but covered by high fixed FiTs for green electricity up to €0.16/kWh for solid biomass. Price for the heat into the grid is €0.02/kWh, price for heat to the consumer is €0.039/kWh and the price of electricity is €0.16/kWh (RENET, 2006).

The initial cost of the gasification power plant was covered by grants, 40 per cent of which came from the Austrian and the Burgenland regional governments. As a member of the EU with an Objective 1 status, an amount of €6 million was made available. To qualify for Objective 1 status the GDP per capita for the region must be below 75 per cent of the EU average (Wikipedia, 2008a). Half of the rest of the money came from fees charged to buildings that connected to the system, and the remainder was borrowed from a bank. It took four years to raise the capital investment of €11 million and attain all the grants and approvals necessary (Douthwaite, 2006). The solar demonstration plants required an investment of €130,000, of which €23,000 was financed by the EU.

Lessons learnt

Güssing's energy plan experienced a variety of challenges since its inception. During the construction of the gasification plant, the EU received a complaint by a local company criticizing the tendering process. Work continued but the EU progress payments were immediately stopped. An unresolved situation for two years resulted in steadily increasing costs to the construction budget. The town's mayor and engineer took the plaintiff to court but before proceedings were concluded, the company had gone bankrupt (Douthwaite, 2006).

The reliance on wood for fuel is a significant issue for the town because only half of the municipality is covered by forests. Plans for more plants (Aichernig et al, 2002) combined with an annual consumption of around 24,000 tons of wood (Douthwaite, 2006) pose a challenge to biomass availability. Some have argued that if only one third of the forests is consumed (RENET, 2006), this would be enough to supply Güssing with all of its electrical power and district heat. The town is conscious about looking after the surrounding forest to ensure they have a good supply of renewable resources for the future. There are plans to grow timber using short rotation forestry methods, which will require some of the arable land previously used for growing maize (Douthwaite, 2006). The switch to solar power will also turn the focus away from biomass (Tirone, 2007).

Today the farmers of Güssing are seen as key energy providers. Most of the farmers report positive feelings about their contribution to the community and now value their work even more. They feel a closer connection to the community. Most of them are on long-term contracts to grow energy crops at a fixed price, which are higher than the market value so they can make a good income. The municipality can thus offer lower energy costs for residents and new businesses (White, 2007). The price is fixed for a duration of ten years, which is currently about €0.016/kWh (Aichernig et al, 2002).

Güssing's energy concept has transformed the region within 15 years into one with a high living standard and low unemployment (BMVIT, 2007). Today Güssing has become an important location for industries with high-energy consumption, such as parquetry production and hardwood drying. The companies use timber coming from local woods and benefit from the cheap heat supply and good market for offcuts (Douthwaite, 2006). A significant coup was the entry of Blue Chip Energy, the first company in Austria to produce high-efficiency solar cells, and Solon AG, one of Germany's first solar energy companies, which reportedly came to Güssing only so they could power their plant with energy from renewable resources (Wikipedia, 2008a).

In 1996, the European Centre for Renewable Energy (EEE) was founded in Güssing to enable companies within the region to share and export their renewable energy technologies and expertise (Alden, 2007). The municipality hosts a team of highly trained technicians and international scientists working to develop innovative technologies, solutions and patents. The biomass plant, for example, has become a research laboratory for studies such as the synthesis of methane, the operation of fuel cells and production of liquid fuels. A pan-European project with participation of Volkswagen, Daimler-Chrysler, Volvo, Renault, BP, EDF and other partners (RENEW) is investigating other biofuel alternatives. Another project is focusing on solar cooling to determine ways the sun can be used as an energy source for air conditioning (Urschick, 2004).

With energy production comes energy management and a research project is underway to determine future cost savings in operational techniques, monitoring and optimization (Urschick, 2004). It is envisaged that the next plant will be 25 per cent cheaper due to the experience gained and the adoption of new technologies, such as examining the potential of unmanned operations (Lutter, 2004).

In 2003, a Solar School was founded to offer operational training for electricians, plumbers, teachers and other professionals in solar energy (EEE, 2009). With its eight research laboratories, it features a demonstration thermal solar plant for hot water, heating and cooling, and a photovoltaic plant generating electricity to run the thermal solar plant. A programme for secondary school students teaches renewable energy practices and technical skills, such as how to use heat pumps, solar cells and solar panels. It includes practical workshops and site inspections of facilities. This learning programme is now running throughout Europe (Fitzgerald, 2008).

Green energy tourism has brought extra income to the town. People are greeted at the entry into town with signs reading 'ökoEnergieland' or 'Eco-Energy Land'. As part of Güssing's tourism marketing, the brand is also expressed as a green water drop symbol. An average of 400 tourists visit Güssing each week. They are as diverse as farmers, investors and politicians, some coming from around the world to gain inspiration (Tirone, 2007). To accommodate this influx of visitors, a new motel has even been built near the EEE, heated and powered exclusively by renewables, hence true to the eco-energy experience (GreenUpAndGo, 2007). Visitors can also benefit from guided tours to the various plants and other joint cultural activities such as the eco-energy marathon, all organized in cooperation with the EEE (BMVIT, 2007).

In essence, the key to Güssing's success was the leadership shown by the mayor and the town council. Fundamental to achieving common energy goals were community involvement, private partnership, research and development coupled with a solid regional commitment.

Jämtland, Sweden

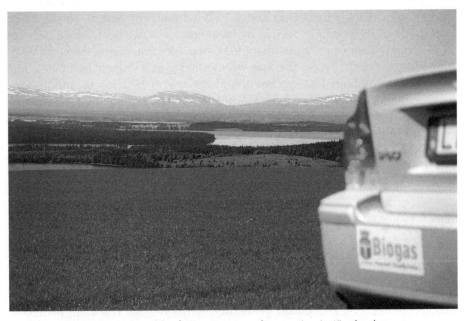

Figure 6.5 Renewables for environmental protection in Jämtland county

Source: Roger Strandberg

Basic facts

Jämtland county is a historical province in the centre of Sweden in northern Europe. It is part of the Mid-Green Belt, an area known for its natural and cultural heritage and high environmental quality. Set in a densely wooded region, with its 128,000 inhabitants, the county accounts for 12 per cent of the total area of Sweden, but only 1.5 per cent of the total population. Jämtland is the second largest province in Sweden, with Östersund its largest and only city in the province as the centre of trade and commerce.

Jämtland is one of the few Swedish regions that was never fully industrialized. Its large agricultural sector traditionally supplied the Norrlandic coast with lumber. Large concentrations of uranium and deposits of, for example, gold, zinc, mica, silver, lead, iron and copper can also be found here. It was not until after the railways were built that Jämtland's economy turned predominantly to tourism (Wikipedia, 2008b).

Jämtland was the first county council in Europe to be certified by ISO 14001 Environmental Management System and EMAS (EU Eco-Management and Audit Scheme). In 2006, Jämtland won the European Solar Prize and joined the EU's four-year campaign on 'Sustainable Energy Europe' (TVLink Europe, 2005). Today more than 60 per cent of all heat and about 90 per cent of all power is derived from renewable sources. The county generates a total of 12.5TWh in electricity, involving 12.3TWh of hydro-power, 0.04TWh of wind power and 0.2TWh of CHP biofuels. Heat production amounts to 2.3TWh, involving 1.1TWh of biofuels, 0.7TWh of electricity, 0.2TWh from heating oil, 0.1TWh from peat and 0.1TWh from heat pumps. In summary, two-thirds of its heating is fuelled by bioenergy or waste heat. Three per cent of the county's transport uses ethanol, as most of its motorists have turned to biofuels. In recent years, the use of thermal solar energy in houses and commercial buildings has increased, and the use of biofuels in middle-scale heating plants (50kW–1MW) also rising. As a result of all these efforts, the county has effectively reduced its CO_2 emissions by 20 per cent within 14 years (since 1990) (Anjevall, 2007).

Jämtland aims to derive all of its energy from renewable sources by 2050, in effect, moving completely away from all fossil and nuclear resources and hence achieve energy autonomy (Eurosolar, 2006).

How it evolved and what drove it

Jämtland is on track to achieve energy self-sufficiency because it has an abundant supply of biomass – the region is almost completely covered by woods and forests. Since the inception of renewable energy in the early 1990s, it has been able, in a relatively short period of time, to advance the use of biomass for heating and power generation in the region. The CHP plant that supply 25,000 households in Östersund, for instance, has made Jämtland's only city one of the most successful in Sweden in reducing CO_2 emissions (Östersunds kommun, 2008). Around 90 per cent of the city's heat and 30 per cent of its energy use is produced in this co-generation plant (TVLink Europe, 2005). The city also collects methane gas from landfills for fuel.

The large amounts of wood residues accumulated during the short summer months are used to fuel a co-generation plant during the long and cold winters. All kinds of wooden materials such as wood chips or sawdust from the local forestry industry or discarded timber from demolition works are burnt in the plant's steam boilers to generate heat and power. For one local, Karl-Erik Strindlund, high winter costs for heating are no longer a problem. In 2003 he converted his heating from electricity to biomass. Hot water pipes buried in his garden supply heat produced in the Östersund plant directly to his house. He has heat exchange units, regulated by a small computer. The new installation has helped him and his family lower their energy bills (TVLink Europe, 2005).

Similar small-scale heating solutions are applied in villages across the county. The village of Brünflo for instance, recently converted from oil-generated heating to wooden pellets and now supplies heat and hot water to 700 households. Produced by local timber companies, the pellets are becoming an increasingly popular clean-heating solution in Jämtland, as well as in the whole of Sweden.

Burning pellets releases as much CO_2 as if the wood had been left to break down naturally in the forest. The pellets are produced from waste woodchips and sawdust that would otherwise be thrown out. Wood pellets are more efficient at producing heat than plain wood because they are dried and compressed. In summer, boilers need only to be filled once a week and in winter, once a day. The traditional wood burning system would in comparison, require regular re-topping. The pellet boilers are quite expensive (around €4500–8500), but most users think it is worth the investment (TVLink Europe, 2005). The boiler provides hot water and keeps the whole house warm. Those who have replaced burning oil and/or electricity with wood pellets have cut heating costs by about SEK500,000 ($64,500) per year and decreased emissions by about 500 tons per year. It is estimated that the average energy demand for heating of a one-family house is about 25MWh each year (Hall, 2003).

There is strong public and political support for wind power in the region. Strömsund in Jämtland, for example, is one of four wind power centres in Sweden. The region has good winds and the proximity to the transmission network will limit power transmission losses (RISI, 2008). Strömsund will be the centre of education for the operation and maintenance technicians dedicated to this industry. It will attract new businesses in the field of wind power, in relation to building infrastructure, services and maintenance and electrical engineering (Hagdahl, 2003) and hence boost employment opportunities.

Jämtland's goal is to ultimately generate 1TWh of wind power by 2015, build distribution systems for alternative fuels and stop the 'single use' of heating oil by 2010. The latest scheme is for a closed working energy cycle involving CHP, ethanol and biofuel production. Peat, timber from local forests and waste sawdust from local saw mills will be used to produce pellets to power the ethanol plant and CHP plant, which in turn produce heat to be fed into the DH system and electricity that powers the pellet plant and sawmills. The pellets are also sold commercially as fuel for heating for buildings not connected to the DH system. Heat is also redirected back and used in the sawmills. The ethanol produced becomes fuel for transport and the methane gas sold to industry and the general public. The organic by-product generated by the CHP plant and ethanol plant becomes fertilizer and is sold to farmers for growing local vegetables and flowers in greenhouses. In sum, the entire process will be a no-waste cycle (Anjevall, 2007).

According to Jimmy Anjevall, Project Leader for the Energy Agency in Östersund, information evenings were very important during first efforts to promote bioenergy. Information had to include not just the use but what investments were required, what installation procedures were involved, and other technical issues regarding DH and heat pumps (TVLink Europe, 2005). The challenge of promoting community participation and re-education was tackled by regular presentations made by Jämtland's Energy Agency about the benefits of renewable energy in all of its eight municipalities. Other forms of promotion included advertisements in newspapers, on TV, radio and the internet, informing residents about electricity use in heating as well as energy efficiency, in particular saving energy in office buildings (Anjevall, 2007).

Overall, the community was supportive. The plan to replace oil heating with DH networks based on biomass in the beginning for instance, was readily adopted, which as a result, led to a reduction in the use of heating oil by 40 per cent since 1997 (Sustainable Energy Europe, 2008). The two following examples show how important promotion was to the financing, implementation and success of bioenergy projects in Jämtland County.

From June 2001 until May 2002, the project 'Comfortable use of wood pellets in one-family houses' was initiated. In cooperation with the Swedish Energy Agency, a regional plan was drawn up with two simple goals. First, the number of one-family houses in Jämtland County heated by wood pellets should increase to 50 and second, the owners of one-family houses would only need to make only one telephone call when deciding to install a wood pellet system. The immediate result of the project was a total of 95 houses using the system (Hall, 2003).

Another project implemented was the 'Thermal solar energy in detached houses' project initiated in 2003. It aimed to increase the interest in thermal solar energy in houses among energy advisors, companies and the general public of Jämtland (Hall, 2004). To foster interest, the first strategy was to coordinate information and education activities for plumbers, installation companies and households. Together with energy advisors to the municipalities, seven information evenings, a regional exhibition attracting 30,000 visitors, and study tours with companies enabled interaction between industry and community (Hall, 2004).

All eight municipalities provided finance for most of the county's bioenergy initiatives, with strong support from local, regional and national politicians as well as industry (Anjevall, 2007). There were grants for transportation, investments and training of staff in efficient driving (Hagdahl, 2003), and incentives and bonus programmes such as free parking for vehicles driving with biofuel (Eurosolar, 2006). Since June 2000, subsidies were provided to stimulate the usage of thermal solar energy in residential buildings. The household solar project, for example, received €20,000 of funding, financed by several private companies, energy advisors and the Swedish Energy Agency, which provided €6000 (Hall, 2004). The agency also supported the household pellet project in cooperation with the Jämtland County Energy Agency (Hall, 2003).

Lessons learnt

The county faced a few challenges when implementing its renewable energy projects. The household pellet project for instance, faced some resistance from traditional plumbing firms when pellet burners were marketed. Information regarding the technology and implementation was unknown and moreover, limited. The common house-owner can make several enquiries into the installation of a pellet system but they were still left with inadequate information. It took several public 'pellets evenings' hosted by companies to finally disseminate the information. And although an increased use of wood pellets has meant an increased production and hence increased employment, the pellets market is not yet ready to operate independently. The pellets market requires more promotion and integration, such as promoting its use in larger scale commercial and public buildings (Hall, 2003).

The household solar project was an educational success and resulted in 11 companies in Jämtland selling and installing thermal solar energy products. However, it failed one major technical target: the installed solar panel area had increased by only 245m², far from its 600m² target. Despite the financial support received from the Swedish Energy Agency, there was still a shortage of institutional subsidy and private investment to lower the cost of panels. Like the wood pellet market, the thermal solar energy market in the region was not ready to stand on it own (Hall, 2004).

The reliance on forests as a primary source of biofuel presents a concern. However, Jämtland in cooperation with the regional companies, municipalities and organizations aims to overcome this by producing 0.2TWh from agriculture in the country by 2015. At present, only 3 per cent of all farmland in Sweden is used for energy crop cultivation. The same applies to Jämtland, which has not been producing fuels from agricultural farming at all. The new goal would entail farmers switching parts of their traditional production to energy crop cultivation (Tyskling, 2008).

Jämtland demonstrates how municipal initiative, drive and commitment aided the development of renewable energy in the county. It also shows how critical public information and education was in improving awareness and strengthening support, through the use of various kinds of marketing and communication strategies. Although some technologies remain costly, government policy has the ability to lessen market constraints and encourage the use of local resources to cover local energy needs at more affordable prices.

Jühnde, Germany

Figure 6.6 Jühnde as Germany's first bioenergy village

Source: Wolfgang Beisert

Basic facts

The small farming community of Jühnde is the first village in Germany to completely replace its fossil energy use for heating and electricity with bioenergy. With a population of 800 people, the village is located in the southern part of Lower Saxony, encompassing 1300 hectares of farmland and 800 hectares of forest. In 2005, the 'Bioenergiedorf' or Bioenergy Village was awarded the European Solar Prize in recognition of its municipal services (Eurosolar, 2005).

Jühnde's renewable energy portfolio consists of a 712kW biogas co-generation plant, a 550kW biomass heating station, two heat storage facilities, boiler and a DH network connecting around 145 houses. Nine thousand cubic metres of liquid manure from six animal farms (800 cows and 1400 pigs), 250 hectares of farmland (25 per cent of total farmland) and 350 tons a year of wood chips from forests (10 per cent of annual growth) are processed to produce bioenergy in Jühnde.

The biogas plant ferments only renewable raw materials in the form of grasses, other agricultural by-products and liquid manure. Within the biogas plant, organic matter is decomposed into methane by bacteria. The gas is then burnt in a block heat plant where the heat is channelled into a bioenergy buffer store. The biogas then fuels a block type power station, generating around 5 million kWh of electricity and 6.5 million kWh of heat energy each year. The biomass plant ensures the heating needs of every household in the village all year round. Around 12,000 tons of non-processed wood, woods with a maximum moisture content of up to 60 per cent and wood chips (Federal Ministry of Economics and Technology, 2008) is processed each year, supplied by seven out of nine local farmers. Fermentation residue is recycled and reused as fertilizer on agricultural cropland in accordance with agricultural regulations (Eurosolar, 2005). The energy crop cultivation involves a double cropping system. By rotating summer and winter cereals, the method helps reduce the use of pesticides and avoid pests and diseases. This works to balance the agricultural ecosystem, generate higher crop yields, ensure diversity and minimal tilling (Ruwisch and Sauer, 2007).

The heat generated by the plants is transported as hot water into the houses via a 5.5km underground network of pipeline. Connected to the circuit of each individual house, the water is fed directly into the heating system. The energy is transferred to the hot water system via a heat exchanger. The process becomes more efficient and beneficial as an increasing number of houses are connected. To date around 70 per cent of households have voluntarily signed heat supply contracts. It costs €49 per MWh, with a €500 fixed annual charge and a €1000 connection fee. Total costs are around €2000 a year (Ruwisch and Sauer, 2007).

Each year the plants generate an income of approximately €1 million, four-fifths from the excess electricity sold and a third from heat energy production. By 2008, the village was producing more than 10 million kWh of electricity, effectively saving 3300 tons of CO_2 annually.

How it evolved and what drove it

The Bioenergy Village began as a research project initiated by Göttingen and Kassel University through the Interdisciplinary Centre for Sustainable Development (IZNE) in 1998. It began with the search for regions with the greatest potential for bioenergy operations – areas with an abundance of biomass from plants and manure, and all forms of otherwise wasted biological material, which could be processed into electrical power and heat. It took two years after its conception before the research project received any financial support.

Four villages were shortlisted following a survey of 3000 households across 17 villages at specially conducted questionnaire workshops. Technical and economic feasibility studies on the short-listed four determined the final selection (Ruwisch and Sauer, 2007). In 2001, Jühnde was chosen as the pilot. It already possessed a large number of farms and 70 per cent of its 200 households were already connected to the hot water grid, the conversion to distributed heating from bioenergy sources would mean immediate CO_2 savings (Bioenergiedorf, 2005).

The planning process with Jühnde began with eight committees encompassing the operating company, biogas plant, energy crop cultivation, biomass conservation, housing technique, central heating plant, heat grid and public relations, overseen by a central planning group consisting of the municipal council, district council, club and associations representatives, church council, university group and other planning groups. When asked what type of operating company the villagers of Jühnde preferred, they unanimously voted for a cooperative. A minimum €1500 investment was then set for all participating households. To date there are 195 resident members and 39 non-resident members (Ruwisch and Sauer, 2007).

In 2005, Jühnde's bioenergy scheme was officially launched following more than four years of preparation. The biomass plant went into service first, following further extensions to the hot water network. The biogas installation went on stream next and by late that year, the village started feeding electricity into the public grid.

The total cost of the project was approximately €5 million. Raised mainly by public funds, 54 per cent came from the German Ministry of Food Agriculture and Consumer Protection (BMELV) and the German Agency of Renewable Resources (FNR) and 26 per cent from state and municipal governments. The remainder was covered by private equity in the form of cooperative shares valued at €0.5 million and bank credits totalling €3.4 million. It is estimated that the biogas facility and co-generation generator cost €2.9 million, heating system €0.9 million and the hot water pipeline €1.6 million. The operating contractor and specialists were required to limit any cost over-run, while maintaining quality practice in planning and implementation (Ruwisch and Sauer, 2007).

Lessons learnt

The first big challenge for the project was to encourage people to disconnect themselves from conventional fossil fuel suppliers and to join the bioenergy system (Ruwisch and Sauer, 2007). This was managed by coordinating planning workshops to educate village groups and to rally their support. The meetings made certain the goals of local residents were incorporated into the project and the right balance was achieved between local knowledge and external guidance. The enthusiasm shown by the village mayor, who was actively involved in promoting the projects, also convinced many residents to sign up (Heiskanen et al, 2007).

The next challenge was the implementation process. There were many delays and steep transaction costs in obtaining planning permissions and construction permits. Frequent revisions of the statutory local plan caused further complications. The biogas plant for instance, took five years from concept to final operations (Binns et al, 2007).

Another challenge was the new role of citizens as energy producers and/or suppliers. Jühnde citizens and energy users had to adapt to being actively involved in the production of renewable energy. This required a period of learning and readjustment as well as some economic and social restructuring within the village (Heiskanen et al, 2007).

The cooperative concept runs on the premise of ecology, economy and social structure as the basis for replacing fossil fuels with locally generated bioenergy. The concept has ensured collective decision-making for the benefit of the entire community, shared responsibility and ultimately shared profits. As a group, the villagers were able to restructure their energy supply and involve themselves in the entire process from planning and management to production and distribution. They agreed that annual costs should not be higher than the costs of a heating system based on fuel oil and insisted that existing and future supplies should be based only on locally grown resources. Fixed prices for energy crops would stabilize long-term agricultural income. Increased output would enable farmers to find new markets for their produce. As a minimum the farmers should earn the amount of money comparable to wheat production (Ruwisch and Sauer, 2007). The selling of wood chips from weak or waste wood, for example, should support local forestry.

For most villagers of Jühnde, communal achievements have been fun (Bioenergiedorf, 2005). The new owner-customers are reportedly very happy because heat produced from bioenergy actually costs less than standard oil fuels. According to the head of the local cooperative behind the plant, Eckhard Fangmeier, the profits alone have been sufficient to run the biogas co-generation plant (Deutsches Generalkonsulat Melbourne, 2005), The projects keep money in the region and have helped improve local knowledge of project management and technical expertise. New jobs have been created through the facilities' planning and construction phases, with more envisaged in the future for operations and maintenance work. Many existing jobs in agricultural and forestry have also been successfully retained (Bioenergiedorf, 2005).

As initiator of the Bioenergy Village model, the IZNE university project group was critical in defining the concept and ensuring the smooth transition to renewable energy sources. It supervised the entire process, conducted progress assessments and motivated community members to become part of the process. The experience with Jühnde has enabled the university team to support other localities with their efforts to utilize renewable energy sources (Bioenergiedorf, 2005).

Jühnde is planning further research into the optimization of its renewable energy processing. A first-trial photovoltaic system is envisaged alongside the establishment of a centre for new energy where people can come and learn. The local authority is consistently looking for partners who would like to use the village as a marketing tool for implementing renewable technologies (Fangmeier, 2008).

The population of Jühnde has increased considerably since becoming a bioenergy village. Some first-time visitors have become new residents, attracted by the idea of becoming part of an energy self-sustaining community. The success of Jühnde has helped refute criticisms directed at pilot projects as irrelevant or a mere curiosity. In fact, it has now inspired over 30 villages in the region to achieve their own energy independence. The daily visits from visitors and farmers from neighbouring villages attest to this.

Mureck, Austria

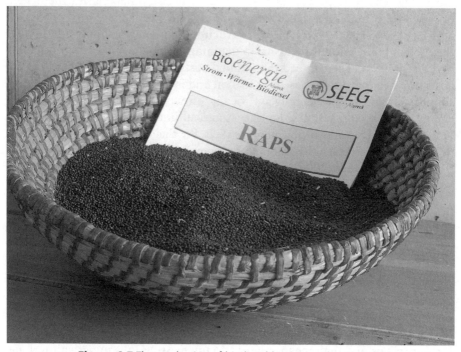

Figure 6.7 The production of biodiesel by rapeseed in Mureck

Source: Greenfleet

Basic facts

Mureck is a small rural municipality in the district of Radkersburg in southern Austria at the border with Slovenia. Its 1700 residents inhabit an area of just 5km². In recent years, Mureck has become a leading model in regional energy self-sufficiency. Its self-sustaining biomass system is one of Europe's first to export surplus energy worldwide (Eurosolar, 2006).

In 2001, its bioenergy cooperative South Styria Cooperative for Energy and Protein Production (SEEG) Mureck won first prize at the World Energy Globe Award for supplying biofuels to the city of Mureck and the wider region. Its achievements gained further recognition with the awarding of the European Solar Prize in 2006. To mark the 20th anniversary of bioenergy in the region, Mureck introduced its sustainable model to the world at the 'First International Climate-Protection-Event' (SEEG, 2007).

Mureck's bioenergy system is the result of the collaboration between SEEG Mureck, Nahwärme Mureck, and Ökostrom Mureck. The three companies form the 'Mureck Energy Circle' or Bioenergy Mureck. SEEG Mureck is an agricultural cooperative belonging to 530 farmer-members predominantly from South Styria, and 70 local authorities. It produces 10 per cent biodiesel from rapeseed and 90 per cent biodiesel from used cooking oil. Farmers cultivate and supply rapeseed to SEEG Mureck for biodiesel production. The farmers later receive biodiesel and rapeseed cake as a high protein feed for livestock. The cooking oil used to produce biodiesel is collected from a cooperative of households and companies. The biodiesel is then returned to the cooperative for use in private vehicles, public buses, freight forwarders, hauliers and petrol stations (SEEG, 2007).

The biomass plant and the co-generation (CHP) plant supply heat to the DH system and are both operated by Nahwärme Mureck. The CHP plant also supplies 'green' electricity. Owned by SEEG Mureck and two farmers, 7.5MW is supplied to around 250 customers, representing 85 per cent of the total heating requirements in Mureck. Approximately 40 per cent of the wood chips used are sourced from 23 farmers and the remainder obtained from sawmills. The cost of electricity generated per household is estimated around €90 including taxes for 1000kWh (SEEG, 2007).

Ökostrom Mureck belongs to Nahwärme Mureck and seven other farmers. Its biogas plant produces 1000kW of electricity by using liquid manure, raw plant materials, maize, silage and glycerine from the biodiesel production. The raw material is harvested within a radius of 6km, collected and fermented in the biogas unit to produce biogas, which is then converted into electricity or heating. An estimated 300 hectares is used for raw materials, with 500 hectares for spreading the biofertilizer produced. It also uses waste heat from Nahwärme's biomass and co-generation plants, so there is no heat loss. The power generated is fed back into the public grid (SEEG, 2007).

How it evolved and what drove it

The switch to bioenergy was conceived by three friends, August Jost, Ferdinand Potzinger and Karl Totter, in a pub in the winter of 1985. The municipality was facing economic problems in agriculture in relation to surpluses in crop production (McCormick and Kåberger, 2007). It was Totter, the present cooperative chairman, who proposed that farmers produce their own source of fuel for use on their property. His farming experience in Austria in the 1950s showed that energy crops could be grown and used again on the same farm (EFFP, 2004). Totter suggested that Mureck should grow rapeseed as the raw material for producing biodiesel and return to crop rotation methods of traditional farming (SEEG, 2007).

From 1987 to 1989, the group began supervising trials at the agricultural college in Silberberg near Leibnitz. SEEG Mureck was founded in 1989 after a year's planning, its biodiesel plant opening in 1991. In 1993, the collection and recycling of used cooking oil commenced, with a specially designed container called 'Fetty' set up for household collections, jointly introduced by the local authorities and waste management associations. Companies and restaurants had to use special containers at fixed sites. Over the next decade, biodiesel production expanded from 0.5 million litres in 1994 to 10 million litres in 2005, which by this time coincided with the opening of the biodiesel filling station. In 1998, Nahwärme Mureck was founded, inaugurating its new biomass plant after two years of preparations. By 2004, Nahwärme's heating network had expanded to cover 200 buildings, reaching a length of 12.5km. Construction on Ökostrom Mureck's biogas plant began in 2004 and opened a year later.

The development of bioenergy in Mureck required substantial funds due to a web of several interconnecting systems. Most of the developments were instigated locally without the aid of a major investor or powerful energy utility. The biodiesel plant in particular was funded from a range of sources. Half of the start-up costs were covered by the Austrian National Investment Fund, with the remaining financed by the members in the form of cooperative shares and external funding received from other agencies and sponsors. The rest of the projects were funded by the EU (EFFP, 2004).

Funding was essential for any of the projects to succeed because of the high upfront costs. The companies have since received between 30 and 75 per cent of the investment costs (McCormick and Kåberger, 2007). Mureck received around €22–26 million, €9.4 million for SEEG, €7.2 million for Nahwärme, €5.6 for Ökostrom, of which up to €1.4 million came from suppliers in Austria and €5–8 million from investments in the supply chain (SEEG, 2007). Support also came in form of access to experts and international research.

Lessons learnt

There have been a few setbacks but none too critical to result in failure. Biodiesel production, for instance, received a lot of opposition from the motor and oil industries. Biodiesel was seen not to work as well as standard fuels. There was very low demand for biofuels and delays in production caused by the low price of fossil fuels. As a non-traditional crop, the introduction of rapeseed also posed new challenges for local farmers.

With regards to used cooking oil as biofuel, availability was very low particularly in the early stages. It took a dioxin scandal in 2000, which banned the use of used cooking oil and fat in animal fodder in Austria, to redirect the flow of used oil towards biofuel production (EFFP, 2004). The potential for cooking oil was also limited by people's average consumption of cooking oil, which was about 3kg per person per year. It was estimated that a catchment radius of 200km is required to be cost-effective (SEEG, 2007)

The large tracts of land required for rapeseed cultivation is a concern for local residents due to the dangers of monoculture. But according to Bioenergy Mureck, it has plans for rotating rapeseed with Sudan grass. A pilot project is researching ways of using complete plants for biofuel that will enable three to four times the amount of liquid biofuel obtained per hectare compared to biodiesel from rapeseed oil (SEEG, 2007).

Rapeseed grown in rotation with other crops in the same area has enabled farmers to keep their fields under continuous production. This rotational method has improved soil structure, prevented erosion, added nutrients to the soils and provided relief to underground water. Weed, disease and insect problems that occur when one species is continuously cropped have also been avoided, so reducing the use and cost of artificial fertilizers (Wikipedia, 2008c).

Converting rapeseed oil into biodiesel does, however, require methanol and energy. Mureck has plans to replace the fossil methanol with methanol made from sustainable sources. The high cost of conversion for vehicles to run on pure vegetable oil (€3000–4000) will lessen with new technology and ubiquity, meeting current conversion costs for standard diesel engines running on biodiesel. The use of pure vegetable oil as biofuel in local farms will also finally take place as costs for third-party drying, storage and handling come on par with the cheaper regional biodiesel system.

The biogas driven co-generation plant in Mureck produces heat and power all year round. However, low heating energy demand in summer has meant that the energy produced during the season is not fully utilized. To use this excess energy, a differential cooling system will be installed via the Nahwärme (close-proximity or neighbourhood heating) flow pipe at 80–90°C, to supply thermal cooling produced by absorption chiller, to retail and commercial establishments as well as detached houses. It will help cool buildings that cannot depend on structural improvements to meet their cooling demands (Beham, 2008).

Bioenergy Mureck is characterized by self-initiative, individual responsibility and close partnerships. It relies on the cooperation of the leading actors as well as good supply-chain coordination to maintain its success (Hametner, 2006). The large constellation of companies and local groups working with the three companies ensures fair representation of local and regional concerns, mix of ownership types and shared accountability that avoids a top-down corporate structure. Majority ownership in the hands of a few was undesirable, particularly when farmers represented only 4 per cent of the population but were the predominant owners (SEEG, 2007).

The systematic integration of all parties in the planning process from the beginning helped quell local misconceptions and instil an understanding of the process. Public information meetings were very important. The local heating scheme for instance, had an immediate take-up of 50 per cent, despite the additional 25 per cent costs of conversion over conventional oil-fired heating systems. As an ongoing service, a customer care hotline was set up as a key link to public energy demands (SEEG, 2007).

Pilot projects were critical in capacity building. Bioenergie Mureck worked with various regional and national bioenergy organizations in Austria to improve its knowledge of bioenergy and share its development experience with them. Among them were the Biomass Association, Energy Association of Styria, Eco-Energy Styria and the universities in Graz. BDI Biodiesel International provided information on biodiesel plant implementation and the cooking oil project received support from waste disposal groups, local councils and restaurants. Partnerships with neighbouring countries such as Slovenia, Germany and Hungary have helped establish new markets for the biodiesel (Hametner, 2006). Mureck has also exported energy to some countries in Asia.

Mureck's bioenergy system has meant energy security, added value and an increase in local jobs. During times of crisis, the municipality can now rely on a secure and more affordable energy supply, which in turn, ensures social stability. By 2006, the annual heating costs for a detached house were €750 less than one equipped with a conventional oil heating, based on an oil price of €0.65 per litre. Today both biodiesel and local heating are far cheaper in Mureck than fossil fuels such as heating oil or diesel (SEEG, 2007).

Bioenergy Mureck is a 'self-sufficient energy loop that can work separately from a conventional supply of energy, an all-round business cycle based on regional energy' (SEEG, 2007). Mureck's bionergy system has created not only new jobs associated with the energy plants such as electricians, machine fitters and joiners for carrying out operational and maintenance work; but it has also maintained existing jobs such as farmers and labourers through production and supply of raw materials on surrounding farms (SEEG, 2007).

According to SEEG Mureck, the town saves 45,000 tons of CO_2 and reduces significant amounts of waste each year. Convenient local heating replacing domestic fires has helped reduce emissions. There are lower emissions from biodiesel compared with fossil diesel in terms of soot, sulfurous oxide and fine particles, though there are some nitrous oxide emissions from engines under some conditions (SEEG, 2007). The biogas plant is fully automated and generates no dust, odour or noise. Separating waste and using mineral oil separator has become mandatory. Biofilters are used to reduce unpleasant smells from the biogas unit and dust filter units are incorporated to reduce dust emissions from the biomass cauldrons. Despite the high cost, textile filter units have been quite effective.

SEEG safeguards the environment by regularly monitoring emissions and compliance with statutory requirements, especially with regard to waste disposal. The SEEG collection system has prevented an estimated 9000 tonnes of used cooking oil from entering the local sewerage system each year, thus protecting the sewerage system and wastewater treatment plants from damage and further expense. According to one study, it costs €0.44 per litre of oil for the upkeep of treatment plants if used oil is poured down the drain (SEEG, 2007).

Bioenergie Mureck takes part in events that promote its goals to the public and industry. Press releases, conference participation and its own website help explain the environmental benefits to a wider audience (EFFP, 2004). Open days provide opportunities for direct discussions with local officials and experts, and ensure good public rapport and public education (Hametner, 2006).

Bioenergie Mureck has significantly promoted tourism in the region. Around 6000 visitors come to see the energy plants each year. In 2001, a visitor's centre 'Energieschauplatz Mureck' on the eastern Styria's Energy Display Road shows visitors how the 'energy cycle' works. The energy concept is also linked to the region's restaurants and local providers of specialities, with joint activities and campaigns regularly staged. There are plans to open an international information, consultation and training centre for renewable energies and industrial raw materials targeting primary, secondary and tertiary schools (SEEG, 2007).

In 2001, a bioenergy beacon was built as a visible symbol of Bioenergie Mureck's four key principles: nature, energy, region and peace, inaugurated on the occasion of the 'Romlauf'. Created by the vocational school in Mureck, the beacon is lit with a fire fuelled by biodiesel provided by SEEG Mureck. As the world's largest lantern, the town's landmark strongly signals the local authority's commitment to the environment, local community and industry.

Mureck local authority has put in place social and economic guidelines in relation to its bioenergy system. Developments must enhance the quality of life for residents, increase the awareness of the region's sustainable cycles, maintain joint setting and implementation of activities, and create jobs, giving people further reason to stay in the region. Maintaining relationships with associations and companies such as schools and the fire brigade and organizing festivals such as the bioenergy festival, biodiesel day and bioenergy beacon have ensured ongoing awareness (SEEG, 2007).

The development of renewable energy in Mureck would not have been possible without local initiative, institutional support, technological know-how and most importantly natural resources. The production of local energy demonstrates how far-reaching the benefits can be, ranging from affordable heating and electricity and increased local income and productivity to improvements in land and air quality and an enhanced sense of local pride and ownership. It demonstrates how private companies in collaboration with municipalities can create socially cohesive, 'greener' environments, while maintaining profits for all.

Navarre, Spain

Basic facts

The region of Navarre is located in north-eastern Spain, between the Basque country and the French border. It consists of 272 municipal areas spread over 10,391km², equivalent to 2 per cent of Spain's total landmass. With a population of just over 600,000 people, it is the least populated of Spain's autonomous regions. Pamplona is the regional capital city.

In 2003, Navarre was praised as having the best regional policy in Europe at the European Conference for Renewable Energy in Berlin. In 2007, a local

Figure 6.8 The National Center of Renewable Energy, Pamplona, Navarre

Source: Oikema

company, Acciona Solar, was awarded the European Solar Prize by the European Association for Renewable Energy (Eurosolar 2007) for its contribution to the development and promotion of solar plants. Today Navarre leads Spain in renewable energy technology and is recognized worldwide for its scientific research and technical capacity.

Navarre's small hydro plants– those with a capacity below 10MW – have been in operation since the end of the 19th century. Today it is no longer a major power source and the potential for new hydro is limited as most riverbeds are well used (Aicher, 2005). To date there are around 111 small hydropower plants in Navarre with a total installed capacity of 195MW, representing 10 per cent of Navarre's energy consumption. El Berbel is Navarre's only large hydroelectric installation generating 18MW.

Navarre's greatest source of energy comes from wind farms. Latest reports account for 32 windparks across 16 locations with 1164 turbines and a total installed capacity of 936MW (Fairless, 2007). Spain's national broadcaster TVE reported that on 19 March 2007, wind power in Spain for the first time exceeded nuclear and coal power's contribution to the power grid. Wind power contributed 8375MW, while nuclear power and coal added 6797 MW and 5081 MW respectively (AP Digital, 2007).

There are around 32 co-generation installations generating a total of 118MW fed by forestry biomass and black liquor. Black liquor is a recycled by-product formed during the pulping of wood in the papermaking industry. Burnt in boilers to produce steam and electricity, the process is effective in recovering inorganic chemicals for recycling. In the town of Sangüesa, a 25MW biomass plant generates enough electricity from 160,000 tons of straw waste to power over 50,000 households, avoiding carbon emissions of 196,000 tons per year. At a cost of €51.9 million, the plant is the only one of its type in Spain (Siehr, 2007). Waste products such as ash are reused as fertilizer.

In Caparroso, a biodiesel plant produces 80 million litres of fuel a year from crude and refined vegetable oils such as rapeseed, sunflower, soya or palm. The €25 million plant was the first in the world equipped to process such a variety of vegetable oils (Gobierno de Navarre, 2006). The fuel is sold to oil companies, freight companies and public transport, and two petrol stations in Navarre that sell 100 per cent pure biodiesel. It is envisaged that the entire city bus fleet in Pamplona will operate with 100 per cent biodiesel, prominently marked with special labels as ecological vehicles. The biodiesel producer has assured the local authority that it will put a cap on fuel cost increases, hence protecting public transport users from price hikes due to the switch to biodiesel (Acciona Energía, 2007).

Navarre's biomethanization plant treats 50,000 tons of urban waste and produces 6,000 tonnes of compost each year. The €9.6 million facility processes the waste to harness methane gas for the generation of electricity. Linked to the solid urban waste plant and wastewater treatment, the system effectively reduces waste sent to landfills.

Recent estimates account for over 600 individual solar photovoltaic facilities working off-grid belonging to private owners. They are used for a range of applications such as lighting, pumping and so forth. Those that are connected to the grid make up the 18 so-called 'solar gardens' spread across Navarre, generating a total 61.5MW and involving more than 3500 people, all at a cost of €456 million worth of investments. A concept created and patented by Spanish company, Acciona Solar, the *huertas solares*® or solar gardens cluster individually-owned photovoltaic installations on a single site (Acciona Energía, 2007).

Navarre's first solar garden was the €12 million installation in Tudela, consisting of 400 solar trackers over 60,000 square metres of land. The 1.2MW garden is divided into two separate zones for energy generation and technology experimentation. Tudela is also earmarked as the site of Spain's first 'zero-emissions' community, balancing energy supply and demand through intelligent energy systems. The eco-city will consist of 70,000 new homes powered exclusively by wind energy and photovoltaics and heated via the use of passive solar techniques, solar collectors and geothermal. The masterplan features bioclimatic architecture as part of its energy efficiency regime (ECO-City, 2009).

In contrast, the 9.55MW Monte Alto Solar Garden in Milagro, south of Pamplona, is the largest garden and generates 14 million kWh a year from 889 automated solar trackers installed over 51 hectares of disused agricultural fields. Owned by more than 750 public and private investors from across Spain, the electricity produced can feed 5000 homes. The solar tracker consists of crystal silicon panels mounted on a structure of 50m² in the form of a grille. The structure is programmed to follow the sun from east to west, depending on the different position of the sun every day of the year, turning on a specified angle to optimize the capture of the sun's rays, which are then transformed into electric power by the photovoltaic cells in the solar panels (Technology Review, 2007).

The headquarters of Acciona on the outskirts of Pamplona is Navarre's first zero-emissions building. All of its energy needs rely exclusively on on-site PV panels, solar water heating, a geothermal system for air conditioning and a small amount of biodiesel. Its architecture aims to reduce energy demand, as displayed in the building's compact shape, use of facade cladding that helps ventilate and shade, and installation of intelligent internal climate control systems. With an energy consumption representing only half of a typical building, the payback period is an estimated ten years (Acciona Energía, 2008).

How it evolved and what drove it

The move towards a comprehensive renewable energy network began in 1989, when Estaban Morrás and two friends founded the Corporación Energía Hidroeléctrica de Navarre (EHN). Their goal was to buy up existing mini-hydrostations, connect them to the grid and build more. However, sites for new hydro were increasingly hard to find. They realized that they could not rely on a single source of energy. Earlier that year, Morrás visited a wind farm in Montpellier, France and discovered that a single turbine could generate as much energy as a couple of small hydro stations. Thus inspired, he approached leading turbine supplier in Denmark, Vestas, to see how he could build his own wind farm in Navarre.

For financial support, EHN turned to the regional government with a wind power development plan. It was not difficult to persuade the government because the plan meant a new industry for the region. While there were obvious environmental benefits, these were not decisive factors. Navarre at the time was heavily reliant on its only industrial employer, a Volkswagen car plant. The region's infrastructure was weak and had limited local power sources, features not appealing to outside investors (Fairless, 2007).

The government began with assessments of the region's wind energy potential. By late 1994, the first windpark was built on the El Perdón mountain range near Pamplona with six 500kW wind turbines. The site had very good wind conditions but was also close enough to be seen from Navarre's largest urban area to ensure a favourable introduction. With the aid of public and private capital, Danish wind technology and further technological developments done locally, Navarre's companies began manufacturing wind turbines of even greater power capacity.

Gamesa Eólica was set up by Sodena, the government's main instrument for local business development and investment, to supply wind turbines to EHN, who by then had become Sodena's leading industrial partner. In early 1996, an Energy Plan was approved by the government of Navarre, which set energy production targets and a regulatory framework for the implementation of wind farms (Ichaso, 2000). In 1997, EHN in partnership with a few local companies formed AESOL to explore the potential for solar energy in the region. AESOL was later sold to Acciona to form Acciona Solar.

In 2000, construction began on the Sangüesa biomass plant as part of the European Thermie Programme and the Spanish Energy Saving and Efficiency Plan and connected to the grid two years later. The Center of Environmental Resources of Navarre was founded by this time to raise public awareness about energy efficiency. In 2002, the Tudela photovoltaic solar energy plant was opened, joining around 1000 photovoltaic installations already connected to the grid. In the same year, Acciona, a major construction and engineering company in Spain, took over EHN.

By 2003, Navarre had more than 88 companies active in various aspects of renewable energy. Two of EHN's previous partners, Iberdrola and Portland, had by this time pulled out because they wanted to continue operating conventional and nuclear power stations, while EHN wanted to exclusively develop renewable energy sources (Korneffel, 2005). In 2004, approvals for new wind farms had ceased. The visual impact on the landscape had become an issue as the government recognized that any further development might lose its public support. In early 2005, the Caparroso biodiesel plant was opened.

By 2006 the Tudela biogas plant and 341 solar thermal facilities were already established. In the following year, the national government began lowering subsidies for wind operators and boosted subsidies for biomass, biofuels and solar photovoltaics (Fairless, 2007). By this time, the Monte Alto solar field in Milagro was connected to the grid and 31 co-generation installations were operating.

From 1995 to 2004, the government of Navarre invested more than €136 million in renewable energy projects, contributing up to 30 per cent of the initial funding and providing tax credits for investors (Fairless, 2007). In 2006 it granted €195.74 million in tax credits for photovoltaic installations alone, with plans for a further €906.77 million in new photovoltaic production over the following years (Gobierno de Navarre, 2007). Private developers also contributed a high level of investment in the launch stages.

Lessons learnt

Navarre's green power plants posed some technical and logistical challenges. The wind turbines for instance, despite large installed capacities, actually generated much less energy due to intermittent wind conditions and maintenance downtime (Fairless, 2007). The retrofit of aging turbines however has helped optimize their effectiveness, avoiding fluctuations on grid due to dips in voltage and thus avoiding the need for new wind farms.

Another technical challenge for Navarre was the region's grid capacity, which has not been able to keep up with increasing energy output (Urien, 2007). Existing grids of 400kV and 220kV are today incapable of distributing the increasing supplies, particularly from wind farms. Grid improvements that guarantee supply and reduce the need for further capital are critical.

While the issue of spare capacity has been partially solved by the installation of biomass and biogas plants, these too have presented difficulties. Burning straw for fuel is difficult due to their high content of chloride and alkaline elements – the result of pesticides and fertilizers used during the growing process. The chemicals produce harmful air pollutants when burnt, corroding the surface of boilers and forming dioxins. On the logistics side, long-term contracts with farmers and cooperatives have been insufficient to guarantee supply and handling of cereal material.

Environmental activists from a local group called Gurelur ('our Land' in the Basque language) fear that there are too many wind turbines and they are not being put in the right place. According to the regional government, wind farms are subject to environmental studies with changes applied to designs before they are authorized. Assessments must be made on the impact of the turbines on bird life (Aicher, 2005). Stringent authorization procedures also regulate the number of wind farms constructed by ruling that new installations should only focus on experimental or self-sustaining systems (Ichaso, 2000). Future windparks must exclude sites of significant natural value and introduce re-vegetation projects for areas affected by building.

Despite its small area, Navarre has become a perfect site to evaluate entire renewable energy systems. Different kinds of solar panels have been tested in relation to such effects as shadows and fog (Technology Review, 2007). The region's growing research and training capacity, especially in cooperation with two local universities, has instilled a sense of reliability and integrity to the network. The National Renewables Centre (CENER) in Navarre is continuing research into the development of commercial applications in wind, solar, bioenergy and bioclimatic architecture, while the Integrated National Center for Training in Renewables (CENIFER) provides assistance in form of occupational training and courses for companies and professionals (Fairless, 2007).

Today a number of companies based in Navarre are leaders in windpark development (Acciona Energía, Eólica Navarra-Grupo Enhol, Gamesa Energía and Iberdrola) and in wind turbine and components manufacture (Acciona Wind-power, Ecotecnia, Gamesa Eólica, Ingeteam and M. Torres). Five thousand new jobs have been created, with over 3000 jobs linked to the wind sector. It is estimated that another 5000 positions in the renewable energy managerial sector will be created by 2010 (Gobierno de Navarre, 2007).

The Navarre energy plan promotes energy efficiency throughout the region through conferences, training courses, energy saving guides and competitions. Public input for various projects was always sought, such as the plan to reduce energy demands in heating and cooling in buildings by installing new insulation, or the plan to reduce electricity consumption in households by replacing old appliances such as freezers and dishwashers with higher energy rated appliances. The latter has resulted in the replacement of more than 8000 obsolete electronic appliances (Salazar, 2008).

The grouping of installations in solar gardens has reduced costs and increased efficiency. There is greater security by being centrally controlled. It allows for techniques such as solar tracking, usually not feasible in independent systems. It also ensures dependable grid connections, sequential components manufacture and the development of software in energy monitoring. Production can be tracked remotely – owners can check their daily, monthly or annual production via the internet (Acciona Energía, 2008).

The model has enabled a larger cross-section of society to invest in solar energy. Previously opportunities to invest in the technology were limited because most Spaniards lived in apartment buildings and shared rooftops. At the Monte Alto Solar Field, an individual panel costs €47,000 but government aid, tax breaks and remuneration for photovoltaic solar energy have helped ease the burden. Shared infrastructure coupled with centralized control of financing, collection of FiTs and facilities management have ensured a zero effect on net income gained as profit for investors (Du Bois, 2007).

The long waiting list of potential investors demonstrates the success of the solar garden concept. As an economic model, the cooperative installations have helped share costs and profits, virtually guaranteeing incomes. Some locals have re-badged the scheme as their 'solar pension fund'. Profits earned off the solar panels they bought today will serve as their retirement income tomorrow (Lungescu, 2007).

The introduction of renewable energy has revitalized many old villages in Navarre. Before the windparks, the villages of Iratxeta and Leoz had only 150 residents. Public services were also very limited. Today the population of both villages has almost double. There is running water, waste collection, new buildings and an influx of new investment thanks to the rent earned from the wind farm operators (Fairless, 2007).

Institutions from several countries have visited Navarre to learn about the region's renewable energy development. Among them were political representatives from the Czech Republic, Hungary, Australia and different regional institutions from Ireland, Romania and Slovakia. A few developing countries are cooperating with the Navarre's government for technical and legal assistance in developing their own renewable energy plans, such as the Dominican Republic with the preparation of its Wind Energy Plan.

The initiative of private enterprise coupled with strong political commitment from the Government of Navarre have been key in the region's success with renewables. The policies articulated in the government's energy plan made possible close partnerships between regional government, private companies and financial institutions. The vertical product structure, from project development and component manufacture through to construction and management, ensured systematic infrastructural support. Navarre shows that local energy supply does not depend on geographic size but on a clever mix of technologies that best harness climatic and landscape conditions. Cooperative ownership has enabled entire populations to actively engage in the production of local energy and improve livelihoods.

Shaanxi Province, China

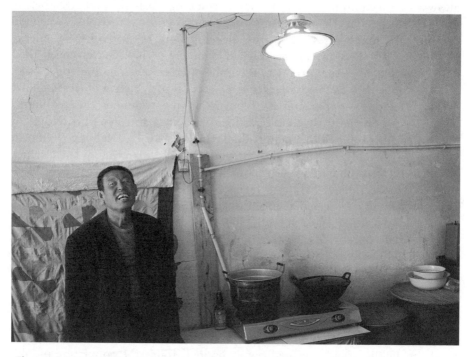

Figure 6.9 Xu Wenjian, Party Secretary, Miao Zuigou Village, Zou Yuan town, near Yan'an, with his biogas light and stove

Source: Ashden Awards/ Martin Wright

Basic facts

Shaanxi is a land-locked province located in central, northern China. Set in a mountainous region, the province is renowned for its natural beauty and historical significance. Xi'an is the administrative capital and the Yellow River its main waterway.

In 2006, a group of women from the region called the Shaanxi Mothers Association won second prize in the International Award for Sustainable

Energy. Their achievement was the installation of almost 1500 biogas plants in farming households across the Shaanxi Province. Thanks to the strong leadership of the group's founder Mrs Wang Ming Ying, the biogas scheme has set forth an energy revolution, transforming countless farms into mini self-sustaining units.

How it evolved and what drove it

The Shaanxi Mother's Environmental Protection Volunteer Association was founded in 1997 with the aim of replanting on the hillsides of surrounding villages. Deforestation (wood was widely used as fuel for cooking) and over-ploughing had caused severe soil erosion resulting in desertification, frequent dust storms and weak springtime river flows. The air quality in the region was poor and the waterways were polluted by animal waste. Living standards were deteriorating with the rise in the cost of food and energy. Recognizing the problem, the local government began restricting tree felling and ordered that hillsides be reforested. Farmers were paid to plant trees and to maintain them. However, the Shaanxi Mothers believed in order to protect the remaining trees as well as reduce poverty, it was essential that an alternative source of energy also be found (Ashden Awards for Sustainable Energy, 2008).

Most households in Shaanxi Province owned pigs and were encouraged by the government to keep them. The local authority offered them one free pig for every pig they bought, with a maximum of two donated pigs per household. The animals produced refuse that leaches into streams causing groundwater pollution. But they also generated a substantial amount of methane gas, which was a valuable source of energy. This resulted in the idea for individually operated biogas systems for rural households. The scheme would be based on a '4-in-1' system, whereby animal, agricultural and human refuse is collected in a digester to generate methane for electricity or as fuel for cooking and heating. The slurry by-product from the methanization process is used as liquid fertilizer spread by farmers over fields or in greenhouses.

To launch the scheme, the Shaanxi Mother's first established contact and gathered support from local agencies and experts in the fields of farming, veterinary science, energy and technology. They held meetings with interested households and village representatives to determine which families could participate and be trained. With the help of technology and agricultural experts, workshops were then organized to educate village volunteers. The programme taught them about the environment and health, the food chain, renewable energy, waste collection, pigsty and toilet retrofit, pipe laying to transmit methane gas, equipment operation, maintenance and repair as well as recovering residue for fertilizers and organic farming (FOE, 2008).

The Shaanxi Mothers promoted solar heaters for water heating and bathing in addition to retrofitting toilets, hence adding value to the overall energy system. Families were also taught how to harvest rainwater. Capacity building workshops were structured so women and their children could attend them together, helping to boost women's self-confidence and encouraging mother–child bonding. In general, villages were involved in the entire process from initial plant assessments to commissioning. Everyone was encouraged to contribute ideas and participate in the decision-making.

Household production of biogas required the installation of a biogas system alongside the retrofitting of the pigsty, latrine and a few household appliances. A biogas digester is first built in a large pit, generally dug by a technician with the assistance from a member of the household. Built from bricks and mortar, this underground cylindrical tank is capped with a concrete dome. Over the digester, the pigsty is built immediately adjacent to the toilet of the house, so pig and human waste can flow directly into the digester. In conditions devoid of oxygen, bacteria begin to digest the slurry. Warm water is added to ensure flow and evenness in solid concentrations. Cow dung is sometimes added for optimal results because it contains the right bacteria. The biogas units are assessed and inspected upon installation. At the end of each year, the units are cleaned and repaired at the same time that gas production drops for two months or so due to the cold winter. Pigsties are meanwhile kept insulated to keep pigs and the biogas digester warm (Wheldon, 2006).

Each biogas system costs around RMB3000–4000 ($400–500) (CunCaoXin, 2008a). Users pay about a third of the cost of the plant, while the remainder is financed by subsidies from the local government and the Shaanxi Mother's. The savings made on fuel and fertilizer within one to two years helps households recover costs. It is estimated that one biogas plant saves RMB600 ($88) per year for coal or wood, RMB250–400 ($37–59) per year for fertilizer and RMB150 ($22) per year for electricity. Combined with the resultant increase in food production due to the residue fertilizer, household incomes can increase by RMB2000 ($293) each year, hence the biogas plant pays for itself within at least a year (Ashden Awards for Sustainable Energy, 2008). Biogas units can last for at least 15 years if there are regularly maintained and used with care.

With an average size of 8–10 cubic metres, a biogas unit produces 380–450 cubic meters of methane gas as daily fuel for an average family of three to five persons for 10–12 months. This amount of fuel is equivalent to 2000 square metres of firewood, RMB600 ($88) worth of reforestation spending, and 200kWh of electricity per household. It also saves on RMB298 ($44) per 1000kg of coal in transport costs (CunCaoXin, 2008a). The biogas is piped into the house with connections to a stove, a lamp or a water heater (Wheldon, 2006).

The installation of units began in 1999 when seven village biogas units were installed in Baota, Yan'an, one of the provincial centres of Shaanxi. By 2001, 153 biogas plants were installed across five villages and within the following four years, another 984 installations were completed in 14 villages. In 2005, Shaanxi's village biogas scheme was renamed the 'Sunflower Project' to help publicize and gain international funding for the work of the Shaanxi Mother's Association. The rebranding occurred with the help of Friends of the

Earth, an environmental charity based in Hong Kong. The sunflower symbolizes the willpower of the rural women because, like the sunflower, they stand upright and proud. They face the sun and a brighter future (FOE, 2008). The Sunflower Project aims to subsidize biogas installations for every rural household in China (CunCaoXin, 2008b).

The Shaanxi Mothers use a funding model based on local government subsidy, philanthropic donations and village household contributions (FOE, 2008). Non-governmental (NGOs) donors include Friends of Nature, Global Women Cooperation, Global Biogas Capacity Building Project, Global Fund for Women (US), Badi Foundation for Village Capacity Building (Macao), the World Bank, WuFang XiaoWei and Friends of the Earth Hong Kong. Foreign governmental bodies, such as the German Consulate, and local government agencies, such as Shaanxi Province Hydrology Office have also provided contributions. Between 2003–2005, over RMB1.2 million ($175,677) was raised. RMB225,000 ($32,940) was spent on capacity building and technological training and RMB915,200 ($133,970) on retrofitting pig sties, toilets and kitchens. Only RMB113,100 ($16,560) was spent on administrative and project management fees.

The Shaanxi Mother's Association and Shaanxi Province Women Federation work together to manage funding expenditure. In general, 60 per cent of funds are paid at the project's launch. The remainder is received once projects are completed and evaluated (FOE, 2008). The village authorities, which include the local party secretary and elected village head, usually decide who should receive financial support. Biogas subsidies are not automatically granted to households that apply because plants are in high demand and funds are limited. Households have to demonstrate how many people would benefit, show that at least four pigs will be kept, and establish that will be enough space to rear them. Users of the unit must also be literate. The announcement of selected households is much cause for celebration and pride. Donor plaques are hung from front doors of recipients. Solar panels installed on roofs become status symbols.

Lessons learnt

The main challenge to the biogas scheme was the limited access to technical guidance and the lack of maintenance experience. Regular education programmes conducted by the Shaanxi Mother's have helped tackle this issue. One initiative was to select 100 farmers who had basic knowledge on biogas management to undergo further training by experts. The intention was to enable farmers to educate their own communities, an initiative that will make certain constant availability of technical service teams in the long run. As an incentive, awards were presented to 20 farmers who performed most outstandingly (CunCaoXin, 2008b).

Many households have found the system to be very convenient and clean. Most women are now able to cook straight after working in the fields because biogas is virtually 'on tap', that is, available at the point of use. They no longer have to go to collect firewood or bamboo from distant forests or buy dirty and expensive coal (Ashden Awards for Sustainable Energy, 2008). Reports suggest that approximately three hours per day was previously spent collecting wood and lighting fires for cooking. This time is now replaced with organic farming, parenting or recreational activities such as weaving and playing music.

According to Mrs Wang Ming Ying, leader and project coordinator of the Shaanxi Mother's Association, biogas is a healthier form of energy (Ashden Awards for Sustainable Energy, 2008). Ever since biogas plants were installed in households, the health of women and their family members has significantly improved. They no longer have to breathe in wood smoke, which was the cause of respiratory and eye disease, because biogas burns with a clean flame. Kitchens no longer accumulate soot. In comparison to those using conventional fuels such as coal and firewood, villages that use biogas reportedly reduced carbon monoxide emissions by 80 per cent, CO_2 by 60 per cent, SO_2 by 80 per cent and dust and fumes by 90 per cent (CunCaoXin, 2008a).

Improved sanitation has meant improved living conditions. The direct connection of latrines to the biogas pit has prevented the spread of odours and bacteria within the home. The biogas digester pit prevents effluent from seeping into the groundwater, hence reducing the risk of waterborne disease through the pooling of dangerous bacteria. The density of flies in houses with biogas has also dropped by more than half, with even significant reductions in pigsties (Ashden Awards for Sustainable Energy, 2008).

Families have benefited from the reuse of slurry residue as liquid fertilizer for crops and fruit trees. As a free natural fertilizer, it has effectively supplanted the use of chemical fertilizers. Furthermore, when crop waste and earthworms are added, a very rich compost is produced that can be readily used or sold (Ashden Awards for Sustainable Energy, 2008). Farmers have reported an improvement between 5 to 10 per cent in plant and fruit growth. According to them, trees are greener and fruits have more colour (FOE, 2008).

Despite facing initial resistance and scepticism, the Shaanxi Mother's Association with its 1200 members, has remained a non-profit organization. It has involved over a million people in environmental and energy projects all over Shaanxi province. With the support of the Shaanxi Provincial Women's Federation (SWF), which belongs to the national network of the All China Women Federation, it has educated more than 14000 rural and urban women and their family members. In recent years, Shaanxi has received many visits from various international delegations such as the US Global Climate Change and Environmental Science Delegation, who were reportedly impressed when shown the extent of biogas use among farmers.

The Shaanxi biogas scheme has facilitated the recovery of the local environment and improved the heath and livelihoods of families. The villages' openness to embrace new technology, entrepreneurial initiative, ability to promote and raise funds and a stubborn drive to overcome bureaucracy has garnered respect from other villages in China. The use of biogas has reduced animal waste pollution and halted severe deforestation. It has increased local incomes based on the sale of organic produce and the availability of cheap energy. More importantly, it has elevated the status of rural women of Shaanxi, who have turned sustainable energy and sustainable farming practices into a form of self-reliance.

Thisted, Denmark

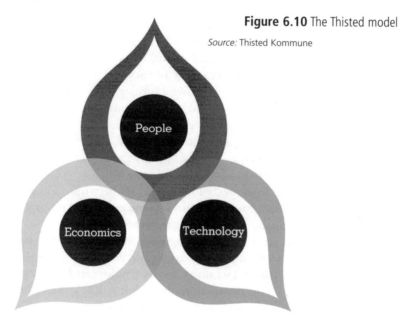

Figure 6.10 The Thisted model

Source: Thisted Kommune

Basic facts

Thisted is a rural municipality located in Northern Denmark, covering an area of 1093km² with a population of 46,000 residents. It is one of Denmark's largest regions, with 100km of coastline and the biggest coastal heathland in Europe. Thisted's main port is gateway to the North Atlantic (Thisted Municipality, 2008a). In 2007 Thisted received the European Solar Prize for its commitment to the advancement, development and integration of renewable energy technologies (Eurosolar, 2008).

Today Thisted generates all of its electricity from renewable energy, 80 per cent of which is generated by 252 wind turbines, representing 103GWh per year. The rest of the energy supply is produced from biogas, solar and power plants that run off residential and agricultural waste. All of its heating is generated from a combination of facilities, which include a geothermal plant, a combined waste incinerator and a straw incinerator. By 2008, around 20,500 households and 1700 companies ran on renewable energy in Thisted (Thisted Municipality, 2008f).

Thisted's geothermal plant is only one of two geothermal power plants existing in Denmark today. The plant produces 25GWh of heating, which constitutes 10 per cent of total heating production. The garbage incinerator generates about 107GWh of heating by processing 52,000 tons of waste a year. This accounts for 70 per cent of total heating demand, also producing 25GWh of electricity a year. The remaining heat production of 30GWh is generated by the straw power plant/incinerator, which processes 8700 tons of straw waste a year, accounting for 20 per cent of heating production. To filter the toxic waste and pollution emitted by burning, wet filters that release only water vapour were installed, an end-process that was a first in Denmark (Thisted Municipality, 2008d)

All forms of organic waste that would otherwise be thrown away are used by farmers in Thisted to produce energy. It is an excellent source of additional income for those who have installed their own biogas plants. For a fee, farmers collect waste product from other local farmers and businesses such as slaughterhouses. The waste is processed in the biogas plant to generate methane, which is burnt to produce steam for heating. Although it may take several years to pay back the biogas plants, farmers get a very good supply of cheap heating. In fact, some are producing surpluses, which at present are lost, because most farms remain disconnected from the DH network. Some farmers have also installed windmills, providing extra income earned off electricity sold to the local electricity company (Thisted Municipality, 2008f).

Recently, a new wood chip based heating plant in Hurup, in the southern part of Thisted, was built to support the existing 9MW woodchip facility. The plant has been nicknamed 'Black Diamond' by its employees. Due to the new facility's ability to process 55,000 tons, an estimated 500,000 kroner ($94,300) is saved each year. An increased storage capacity has also meant minimizing the transport of biomass though town (Thisted Municipality, 2008f).

How it evolved and what drove it

The shift to renewable energy began in 1982 when the municipality founded the Nordic Folkecenter for Renewable Energy. As a non-profit independent organization, it was to provide research and training into the development and implementation of renewable energy technologies and energy savings in Denmark. Funded by local authorities, national and international agencies and industry, the goal was to completely replace fossil fuels and nuclear power with renewable energies (Thisted Municipality, 2008f).

By 1984, the municipality had built its first geothermal plant, also the first of its kind in Denmark. A second plant was built in 1991 to meet the increasing heating demands. The straw-fuelled incineration plant was built in 2005 providing supplemental heating. In 1998, a problem was identified with the energy system. No financial assistance existed for replacing old, outdated windmills with new turbines, a problem that remained unresolved until 2001 (Thisted Municipality, 2008a).

Thisted developed a renewable energy concept called 'The Thisted Model', which highlighted community, technology and economics as key ingredients for sustainable development in the region. The model would ensure cooperation and active engagement by all stakeholders in implementing existing and new energy technologies that also made sound economic sense. Achieving the right balance would more likely prompt politicians, companies and residents to take initiative, encouraging them to participate in the process voluntarily. The model aims to better respond to local concerns and enable a local sense of ownership (Thisted Municipality, 2008f).

Different marketing techniques were employed to promote the Thisted model. Branding was especially crucial during the implementation process and when educating the public. In September 2008, a website was launched showcasing the municipality as the leading climate municipality of Denmark. The site features an 18-minute long movie, explaining in detail the Thisted model. Regular town meetings provide locals opportunities for debate, brainstorming and concern-raising. A community 'ideas bank' was even set up by the municipality for future reference (Thisted Municipality, 2008e).

Thisted has adopted strict energy efficiency standards in residential construction. Recently, a new housing project in Vejle called 'Komforthusene' was commissioned to demonstrate passive houses that only use 9kWh/m2, compared to 58kWh as dictated by Danish regulations, resulting in the reduction of CO_2 emissions by two tons per year. The design involves thick insulation (40–60cm in walls, ceiling and floors), with a low energy air-conditioner coupled with a ground heating system, installed to keep the house warm during winter (Thisted Municipality, 2008g).

Thisted constantly evaluates all its heating and electricity operations, as well as its transportation and building stock. Research is underway to examine ways standard practices could be made more energy efficient. One simple example was staggering school start times in order to use the same school bus travelling to different schools, an initiative that has helped reduce CO_2 emissions from transport by 10 per cent (Thisted Municipality, 2008f).

In the beginning, there were no government subsidies for renewable energy installations in Thisted. Financial assistance only came into existence after the municipality won the award in 1992 for being the first authority to develop wind installations that served its entire region (Thisted Municipality, 2008f).

By 2011, Thisted aims to have increased its electricity generation with 150MW land-based wind power and in the following year, with 400MW sea-based wind power. A wind turbine plan will involve around 80 new, more energy-efficient, more profitable turbines spread across 26 projects. The turbines will be larger, with some spanning 136 metres in height, five of which will be able to produce enough energy for more than 12,000 private buildings. Energy-efficiency measures will be enforced targeting the reduction of energy consumption in new buildings and energy tagging for all public buildings. There will be a comprehensive plan for public transport services as well as for water, heating and electricity. A plan will also look at extending the DH network to include smaller DH heating facilities that supply remote areas. In cooperation with the Nature Conservation and Environmental Organisation, Thisted aims to reduce CO_2 emissions by up to 3 per cent each year until 2025 (Thisted Municipality, 2008f).

To achieve these goals, incentives will be provided, such as compensation for people who live very close to wind turbines, renumeration increases for wind energy and biogas, subsidies for solar and wave energy and tax-free provisions for hydrogen and electric cars. Incentives will be consistently re-evaluated depending on the energy system in operation, community needs and economic benefits (Thisted Municipality, 2008a).

Lessons learnt

There have been many challenges faced and lessons learnt. One challenge was achieving equity in energy distribution in terms of energy production and delivery (Thisted Municipality, 2008f). Lack of financial incentives, incomplete energy infrastructure and distances were the other major barriers.

Despite this, the switch to renewable energy was not difficult for some because many farmers in the municipality already possessed windmills and operated their own biogas plants. However, these plants must merge with the DH network soon in order for farmers to sell surplus heat to other customers in the future. Indeed, a more complete renewable energy grid will help deliver surpluses to the entire network and hence provide farmers with more income, as an increasing number will start to explore waste collection and energy production opportunities. The extension of the network will also mean strengthening rural settlements, as they become self-sufficient and enjoy cheaper energy supplies. It is estimated that energy costs in the region has decreased by two-thirds since the use of renewables in comparison to bills based on oil use for heating (Thisted Municipality, 2008f).

Today Thisted Municipality promotes itself as an environmentally sustainable region and leader in renewable energy technology. The Nordic Folkecenter for Renewable Energy in the municipality has become one of the biggest eco-sites in Europe. Thisted's positioning to be leading CO_2 neutral municipality in Denmark and leader in CO_2 reduction has already attracted visitors not only from Denmark but also from all over the world (Thisted Municipality, 2008b).

The conversion to renewable energy without local government funding was only made possible by the strong support and commitment from local communities and businesses. Its success can be attributed to the intelligent use of technologies, combined with local business investment and citizen enterprise. The use of renewable energy in the area has attracted new companies to the area and as result, increased employment. New businesses based on waste collection, energy generation and delivery have flourished. Thisted municipality believes that the reduction of CO_2 emissions no longer depends on heavy investment in advanced technology. A rethinking of standard procedures based on some lateral thinking is sometimes sufficient.

Initiatives were implemented to boost tourism and increase environmental awareness. Organic farming was encouraged to rehabilitate the agriculture sector. Acting as promoter and steward, the municipality assembled a consortium of pertinent parties, which included the public utilities company, ACAM; the regional, provincial and Mountain Community authorities; breeding and farming cooperatives and a regional energy agency called ARE Liguria Spa (IDAE, 2001b)

Two wind turbines were first installed, with a further two in recent years. The total cost was €1.8 million, 30 per cent of which was financed by the EU and regional funds, with private investments constituting the rest. The installation of solar panels on municipal roofs soon followed. These were funded by regional and local funds, amounting to €155,000 (Procura, 2008). To register for the ISO 14001 and EMAS, local funds of €51,000 were provided. The EU and Mountain Community provided some financial assistance to promote organic farming. For its municipal waste management, which includes collection separation and landfill site management, local and regional groups raised €32,0000 (Green Labels Purchase, 2006).

Community participation in renewable energy projects was encouraged. The municipality began running a school project called the Force Énergétique par les Enfants (FEE), as part of the EU programme. The project raised the awareness of local students, families and local stakeholders on energy issues such as energy saving, renewable energy sources and the environment (Procura, 2008).

Lessons learnt

According to the town mayor, the main obstacle to 100 per cent self-sufficiency in rural Italian villages was bureaucracy. The high connection costs to the national grid and general lack of funding were major challenges. Varese Ligure had to wait one year to get approval from Italy's central authorities for its wind farm project and another year to connect to the national grid. The project finally cost half of the €900,000 originally quoted by Italy's main electricity supplier ENEL. The small hydroelectric dam, which was to produce 1 million kilowatts at a cost of cost €1.1 million, has been on hold for months because the village has been unable to obtain funding. At an average price of €20,000, solar panels were also very costly. There were no grants available to private citizens and businesses wanting to install them.

However, according to the national government, there will be measures to increase the use of renewable energy. These include tax rebates on solar panels, funding of all small-scale renewable energy projects, incentives for purchasing energy-efficient household appliances and industrial equipment, and a €25 million fund to finance education, training, information and international cooperation each year. Modelled on a similar law in Germany, an 'energy account' law will be established as part of a system of incentives. The law authorizes electricity suppliers to sign 20-year contracts with solar panel owners under which all the energy produced by the panels is sold to the supplier at a price about two and a half times the electricity sold by the supplier to private consumers. The extra kilowatt hours are then deducted from the solar panel owner's electricity bill.

Because of Varese Ligure's rural landscape and sparse population, renewable energy technologies were well-suited to provide cheap, more accessible energy to a region, which would otherwise be disconnected due to large distances from major transmission lines. This is despite the mixed reactions to the wind farms, as some see them as a disfigurement while others find them oddly striking. Still, the large amount of surplus electricity has meant extra income for the municipality and low energy costs for locals. Also, the 50 agricultural cooperatives have reported very good profits.

For locals, the strategy adopted by Varese Ligure has meant economic progress, energy security and higher living standards. Within ten years, 140 new jobs have been created within the renewable energy sector. Most people are staying, former residents have returned and some newcomers are coming to live in the area, attracted by the clean air, clean streets and organic produce. It has also meant environmental and historical preservation for the municipality. As well as a large supply of cheap energy, Varese Ligure's council receives an additional €350,000 ($514,000) in tax revenues each year from the private company that owns the renewable energy network. According to town mayor, Varese Ligure easily fulfils all the requirements of the Kyoto Protocol. Any profit that is generated goes back into the local renewable energy operations, thus keeping energy costs down for residents (Burgermeister, 2007).

EU policy was key to the development of renewable energy in Varese Ligure. The EU had set definite objectives in relation to renewable energy development that, in accordance with financial agreements, Varese Ligure had to fulfil (Burgermeister, 2007). Also instrumental was local political drive, determination and patience, particularly in relation to approval and funding periods. Acceptance and participation by locals ensured fluent implementation.

The number of tourists visiting Varese Ligure has increased significantly in the last ten years, many arriving just to see its renewable energy facilities (Thomas, 2008). The ISO 14001 and EMAS certifications have helped raise environmental awareness in the village, increased the standing of the village in the region, as well as promote renewables as a viable alternative to other villages. Carbon emissions reductions have been significant. The electricity from wind turbines alone has reduced carbon emissions by 8000 tonnes (Burgermeister, 2007). Most of the public buildings in the municipality are completely self-powering and its food supply is virtually guaranteed by the many organic farms at hand.

The steely determination and resolve of the rural council of Varese Ligure has been instrumental in the development of renewable energy in the region. By exploring the various financial mechanisms and incentives at hand, it was able to fund a locally based model that regenerated whole villages, conserved history and combined local energy production with organic food production as a robust basis for sustainable tourism.

Eco-Viikki, Helsinki, Finland

Figure 6.12 Solar panels are integrated to a block of flats in Latokartano ecological housing area in Viikki area, Helsinki, Finland

Source: Pöllö

Basic facts

Eco-Viikki is an ecological experimental area in the southern part of the Latokartano housing area in Viikki, 7km northeast from the city centre of Helsinki in Finland. This eco-community project is spread over a total area of 23 hectares on a greenfield site, bounded by a historical nature conservation area and the University of Helsinki's School of Agriculture and Forestry. Today, Eco-Viikki is known for its high-quality high-density ecological housing construction that incorporates intelligent energy systems alongside social aspects of sustainability. Today, some 1700 inhabitants live here and enjoy many conventional services such as supermarkets, clubs, bus stations, a kindergarten and schools (Energie-Cités, 2000).

Eco-Viikki employs new technologies in energy production and savings. With regard to solar energy, Eco-Viikki applies passive techniques involving building orientation, green houses and glazed balconies, while active techniques involve solar-generated heating and electricity systems. Tests on photovoltaic cells integrated into multi-family buildings have shown a production of 15–20 per cent of the household's needs from just 200m² of photovoltaic cells built into the balcony railings. The solar systems generate approximately 80–100kWh/m², with a capacity of 24kWp. They are connected to the grid, which allow surplus electricity to be exported in summer and the import of energy during the colder winter months (Intelligent Energy Europe, 2008).

Eco-Viikki's solar heat project was the largest solar energy installation and first ever project of its kind in Finland. Today, there are nine solar energy systems of between 80–250m², each optimized for its integration into roof construction and canopies, with a total collecting surface area of 1246m², installed for 368 apartments (Energie-Cités, 2000). The solar heating systems are utilized mostly for the heating of domestic hot water and also as a supplement to space heating through floor heating. New solar combi-systems, low-flow schemes, parallel use of solar and DH and large area modules (unit size 10m²) are all demonstrated in Eco-Viikki (SOLPROS, 2003).

How it evolved and what drove it

Eco-Viikki began as an eco-community project, a collaborative project established in 1993 by the Ministry of Environment and the Finnish Association of Architects (SAFA) with the aim of testing 'ecological principles in practical design and building' (City of Helsinki, 2005). In January 1994, expressions of interest were sought from local authorities throughout Finland for a testing ground for the project. From 16 proposals, the area of Viikki was chosen as the site due to its proximity to the city centre, accessibility via public transport, existing communal structure and because preliminary plans to settle Viikki were already been in place since 1989. The local plan envisaged the extension of the university area and the construction of a new residential area connected to Viikki's Science Park, a plan that conserves the natural and cultural values of the area (Intelligent Energy Europe, 2008).

There was a huge amount of interest when the City of Helsinki, the National Technology Agency of Finland (Tekes) and the eco-community project launched the double competition in 1994 for the planning and design of Eco-Viikki. The first competition concerned the layout of the district and the second targeted the building of the experimental housing community. Competition criteria stipulated that the design should be economically feasible, possess high architectural quality, be integrated well with the environment, be pleasant to live in and promote diversity of uses and cultures. The design should also integrate the use of renewable energy, incorporate water-saving measures, maintain ecological diversity and the use of durable raw, non-toxic materials whenever possible. Other criteria included implementing modern telecommunication technology, promoting the involvement of residents in environmental protection and integrating solutions without cars, with priority given to public transport (Energie-Cités, 2000).

The winning proposal was chosen in 1995 from 91 proposals. The design by architect Petri Laaksonen was based on a 'green fingers' concept. Open space between buildings allowed for planting and allotment gardens, composting and storm water management. Mainly four to five stories in height, residential buildings gathered around courtyards with facades oriented to the south to maximize solar exposure. Stepped building massing and vegetation lessened the effects of wind through open spaces (WSA, 2004). Every housing development included a separate metering of water consumption for each dwelling and water-saving plumbing fittings.(City of Helsinki, 2005). Achieving an estimated consumption that was well below the competition criteria, the architectural design proposed careful use of ecological materials and renewable energy use in form of solar heating (Energie-Cités, 2000).

By 1997, a set of ecological criteria (PIMWAG) had been drawn up by an interdisciplinary working group to evaluate subsequent building competition proposals for different lots. The eco-criteria defined five aspects: pollution, natural resources, health, bio-diversity and growing food (WSA, 2004). No specific requirements were made in relation to building construction, but the buildings were required to achieve a high standard in baseline performance. Calculations and explanations were required in the building permit documents and signed by the developer to show that mandatory requirements were fulfilled (Intelligent Energy Europe, 2008).

The first phase of Eco-Viikki began in 1998 with construction of a third of the district. By the spring of that year, a methodology that integrated monitoring and assessment of the projects was published (Energie-Cités, 2000). By 1998, Eco-Viikki was completed with 2000 inhabitants moving in, generating around 2000 new jobs (WSA, 2004). The Eco-Viikki solar project was launched in the summer of 1998, but it was only in 2000 when the first solar system was installed with a total surface area of 248m². By late 2001, solar systems were contributing some 15 per cent of the annual heat demand of the Eco-Viikki area (Energie-Cités, 2000). The new Science Park is due for completion in 2010 as an international centre for biology and biotechnology of the University of Helsinki, creating around 6000 new jobs, 6000 places for student and homes for 13,000 people (WSA, 2004). The remaining residential areas of Viikki will be constructed after 2010.

Eco-Viikki evolved as a response to the Rio Climate Change Conference and Kyoto Agreement. The domestic driver was the Finnish government's programme of ecologically sustainable development, which promoted a cut in energy use (WSA, 2004). Through the KEKO programme, Tekes was able to test the principles of sustainable development in eco-construction by supporting the development. Ecological models from other Nordic countries were also influential, particularly during the earlier stages (City of Helsinki, 2005). SOLPROS, an independent Finnish company acting as primary consultant in relation to energy, environmental and health issues, coordinated the entire solar project as part of the EU's Thermie Programme (SOLPROS,

2003).

The Housing Fund of Finland Housing provided major finance for the new district. To compensate for the cost of building foundations on difficult clay soil, the City of Helsinki reduced land rents. The €4 million subsidy granted by the EU was channelled primarily into research and product development projects such as building materials and methods, environmental management and waste management; rather than actual building and infrastructure construction (Intelligent Energy Europe, 2008).

Lessons learnt

Many of the residential innovations that were initially proposed were not implemented. For example, proposals for using clay and recycled building waste for floor structures, recycling grey water and cooling cold storage spaces by using geothermal energy failed due to technical and economic reasons. Thermal concrete wall construction was abandoned due to problems encountered during construction. This stage also caused some builders to forego restrictions on forbidden building materials as stipulated in the ecological materials database. The information network planned for buildings and the whole area did not proceed because broadband connections from external operators were already commonplace (City of Helsinki, 2005).

Despite the diversity of housing types, high density was an issue for some residents because they expected to be closer to nature, as the name Eco-Viikki suggests (City of Helsinki, 2005). Ironically, it is this density that enables the preservation of nature in surrounding areas. Nonetheless, the project has attracted many environmentally conscious consumers (City of Helsinki, 2005). When marketing the project to future purchasers and tenants, the ecological features of the district were key selling points.

The City of Helsinki was ultimately responsible for initiating and developing the Eco-Viikki area. From the very beginning, the utilization of solar energy was enforced in the preparation of building and land use plans for Eco-Viikki (WSA, 2004). At a cost of €800,000, the Eco-Viikki Solar Project was unique because solar energy was integrated into the project at different levels, using roof-integrated collectors and solar design, among other solutions and new financing patterns such as joint purchases (Energie-Cités, 2000). Because building planning did not start before the solar project, it allowed for the proper siting, placing and sizing of solar heating systems in combination with proper building orientation. The building planning process therefore encompassed all solar requirements, both active and passive.

The utilization of solar energy and natural ventilation techniques has been quite successful (City of Helsinki, 2005). The solar heating storage system fits well with the basic heating system, which is DH with low return temperatures, which means that the solar heating system can produce higher solar yields than in standard systems. Although the solar collectors have functioned well, the systems' heat discharge circuits still required adjustments after its first year.

The property maintenance staff had to be trained in the use of the equipment (City of Helsinki, 2005). The cost of solar installations has increased house prices by around 0.5 per cent but residents do receive free hot water in return (Energie-Cités, 2000).

Despite heat energy savings of around 25 per cent, the goal of using 33 per cent less DH than conventional Helsinki residential buildings was not achieved during the first years of operation, in fact actual consumption rose by almost 15 per cent. Factors influencing heating consumption included ventilation, house and ownership type, solar heating and household hot water use. The actual electricity consumption in Eco-Viikki has varied even more than heating, as a result of similar factors plus the use of additional household appliances, individual saunas in flats, the amount of shared spaces and lifts (City of Helsinki, 2005).

Nonetheless, the consumption of primary energy, that is energy bound up in materials, has been reduced by one fifth, compared to conventional building (WSA, 2004). There were reductions in the amount of waste produced by an inhabitant, the CO_2 emissions of each building and the amount of building waste during construction (Intelligent Energy Europe, 2008). Water consumption decreased by 22 per cent, but the figure also varied depending on the house type and form of ownership.

To maintain control over development and promote environmental quality, a land exchange agreement stipulated that one-third of the plot area would belong to the Finnish state, reserved for building owner-occupier housing. The City of Helsinki and the state would then jointly agree on the plot subdivision and the form of occupancy, the divided plots then subsequently sold to building developers based on contracts dictating overall ecological objectives (City of Helsinki, 2005). About half of the housing was reserved for owner-occupancy and a quarter each for rented and 'right-of-occupancy' dwellings (enabling residents to buy into the scheme by paying a refundable amount of 15 per cent of the value of the home, alongside a monthly residence charge). It was easier to implement additional funds to fulfil eco-criteria with owner-occupied housing than in state-subsidized housing (City of Helsinki, 2005).

The competitions were critical in the development of the project, particularly with awareness raising and the training of design and construction teams (Energie-Cités, 2000). The competition structure insisted on workgroups that included a variety of professionals. The range of expertise was necessary to ensure that projects were successfully executed, operated within a realistic budget and fulfilled all eco-criteria originally proposed. Each workgroup had to include an architect, a structural engineer, a HVAC (heating, ventilation and air-conditioning) and electrical engineer, an expert in ecology and a building developer (City of Helsinki, 2005).

Finland is a country where there is strong urbanization so Eco-Viikki is set to prove that integrating ecological criteria in urban development projects is feasible. A holistic system from beginning to start based on ecological development principles is possible but it requires patience, dedication and many years of collaboration before the benefits come to fruition. Although Eco-Viikki is not yet complete, the project has shown the world that much could be accomplished. Environmental evaluation of the schemes and many technical innovations in construction will provide important areas for future learning.

Woking, UK

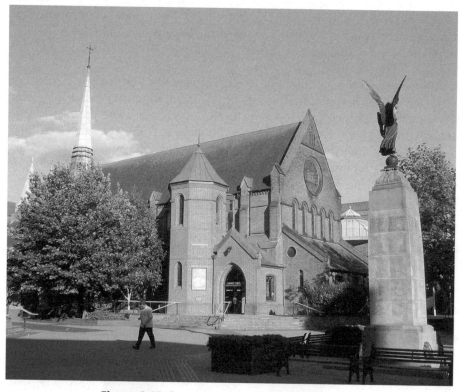

Figure 6.13 The town square of Woking, Surrey, UK

Source: Alan Ford

Basic facts

Woking Borough Council is located in Surrey, 30 miles south of London. The borough has a population of around 100,000 people and is surrounded entirely by greenbelt land, including large areas of countryside, heathland and common land. H. G. Wells was its most famous resident and wrote *War of the Worlds* here (Brown, 2004). Today Woking is recognized as the most energy efficient local authority in the UK and is renowned worldwide for its local decentralized energy systems.

In the millennium New Year's honours, Allan Jones, the council's then energy services manager, received an MBE (Member of the British Empire) for his services to energy and water efficiency (Muir, 2005). In the following year, Woking was awarded the Queen's Award for Enterprise in Sustainable Development for the development of local sustainable energy systems, the only local authority to ever receive the prize. In March 2005, Woking was one of seven local authorities awarded the Beacon Council status for Sustainable Energy. In partnership with the Energy Centre for Sustainable Communities, the status promoted Woking as mentor to other local authorities.

Woking's energy services concept is not about the provision of electricity and gas but energy services such as heating, cooling, lighting and power (Woking Borough Council, 2003). Woking was the first local authority in the UK to introduce the ESCO model, a public–private joint venture energy services company, ultimately responsible for a network of over 85 local generators to directly power, heat and cool municipal buildings, residential estates and local businesses. The distributed small-scale power plants range from 3–10,000kW and are located close to where electricity is used as an alternative source of energy or as supplement to existing conventional supplies (C40 Cities, 2008a).

Woking's town centre energy plant incorporates the UK's first small-scale CHP heating and heat-fired absorption cooling system. It produces 1.3MW of electricity, 1.6MW of heating and 1.2MW absorption cooling (Morgan and Thorp, 2005). The borough's private wire residential CHP systems were the first in the UK, and its integrated photovoltaic installations were one of the largest in the country. Car parks have solar-powered ticket machines, estates have roofs covered by photovoltaic panels and streetlights are either solar or wind powered (Brown, 2004). A 34m-long, 22m-wide canopy covered with 35,000 photovoltaic cells creates a spectacular entry to Albion Square train station, marking the gateway into town. It will produce up to 58MW of electricity a year, saving 41 tons of CO_2 emissions. The solar photovoltaic panels and CHP have been particularly beneficial in the council's public housing schemes (Stotz, 2005). When flows of electricity from photovoltaic cells stop as daylight ends, the CHP plant kicks in. When rooms in town are getting too warm, excess heat is used to drive the chillers.

Woking was the first local authority to install a hydrogen fuel cell CHP system in Woking Park. It produces 1.1MW of electricity, 0.2MW of electricity by hydrogen fuel cell, 1.6MW of heating and 0.5MW of absorption cooling. Excess heat produced is used to power the leisure centre's air conditioning requirements via heat-fired absorption cooling. The fuel cell plant also supplies pure water as a by-product of hydrogen and oxygen combining to produce electricity, producing roughly 1 million litres a year of 100 per cent pure water (Brown, 2004). Surplus electricity, representing 60 per cent of the energy generated, is exported to the council's sheltered housing schemes (Woking Borough Council, 2003), providing power as well as water to two hotels, a conference and events centre, a bowling alley, a nightclub and a multi-storey car park (Muir, 2005).

Woking implements a range of low carbon projects including energy and water efficiency, waste recycling, energy from waste and alternative fuels for transport. To reduce emissions, the council has replaced its diesel-engine refuse collection lorries with ones that run on liquefied natural gas (Muir, 2005). The replacement occurred progressively as transport contracts came up for renewal. Householders were offered condensing central heating and water boilers for the same or lower price than a conventional boiler as part of an energy conservation package (Stotz, 2005).

How it evolved and what drove it

Woking's decentralized energy transformation began in 1990 when Allan Jones, the energy manager for Woking Borough Council completed an environmental audit reporting on global warming, which swiftly resulted in an energy efficiency plan established later that year. Incidentally, the Jones' report came two years before the Rio Earth Summit. Their goal was threefold: reduce CO_2 emissions, adapt to climate change and promote sustainable development. With this in mind, the council began revising their planning and regulations, energy services, waste, transport, procurement, education and promotion, management of natural habitats, and adaptation measures to climate change (Morgan and Thorp 2005).

Woking Council first established environmental and energy targets without knowing how much it would cost. It was agreed that politics had to be put aside before any financial proposals were presented. However, it was soon estimated that £1.25 million was needed to reduce emissions by 20 per cent over five years, which was a huge initial capital outlay (Stotz, 2005). So in lieu of a lump sum, its energy manager asked for a fifth of the money needed for the five-year target, with a proviso that financial savings or profits made from the reduced energy bills would be recycled into the following year's investment pool. Indeed, a year later, the council was already saving around £700,000 on its energy bills (ABC Radio National, 2008). The energy efficiency recycling fund helped change attitudes within the council. The large savings achieved resulted in an increase of bi-partisan support among councillors for the Energy Efficiency Policy (Stotz, 2005).

By 1991 Woking had launched an energy-efficient lighting system for municipal buildings, which generated savings of up to 70 per cent. This was followed by the installation of the £4.2 million co-generation system and a Building Energy Management System (Stotz, 2005). In 1999, Woking Council founded its own energy and environmental service company (ESCO) called Thamesway Ltd. To be free of local government regulation and restrictions in government capital distribution, Thamesway Ltd created its own unregulated energy services company called Thamesway Energy Ltd, predominantly owned by the borough council with 10 per cent ownership by Danish company, Xergi Limited. The company and the council's revolving fund helped finance the installation and maintenance costs of the CHP plant. Typically, Thamesway

Energy Ltd covered the net capital costs, with running costs met by income received from tenants (Stotz, 2005).

Most households voluntarily signed up for renewable energy. By this time, Woking had already implemented 10 per cent of UK's photovoltaics as well as put into operation its first fuel cell CHP. Cables were laid as part of the private wire network. In 2003, the fuel cell plant at Woking Park was launched. By 2004, the Surrey Structure Plan required that all development incorporate energy efficiency best practice procedures. A minimum of 10 per cent of the energy should come from renewable resources generated on-site. CHP for over 5000m$_2$ of development was encouraged (Morgan, 2006).

The ESCO finance model helped progress and maintain the recycled capital fund. Within nine years, a total of £2.2 million was invested in over 85 projects, from the original capital fund of £0.25 million established in 1990. An 8 per cent internal rate of return with the council and green energy sales based on market rates ensured decent income on investments (C40 Cities, 2008a). By the end of 2006, the council's energy company (Thameswey Energy Ltd) had invested £12 million.

Some of the council's assets bought for £3 million were developed between 1991 and 2000, and another £9 million (2000–2005) was spent building the council's new infrastructure. The electricity network itself was relatively cheap. The main cost was the CHP system itself and the DH network. Trenches were already in place when DH was put in, so the cost of dropping a cable for the delivery of electricity, heating or cooling in the same trench was negligible (ABC Radio National, 2008).

Woking's energy efficiency strategy was financed by a £256,000 grant-aided energy conservation scheme provided by the council and supplemented by the national government. This helped provide full insulation measures in over 3000 fuel-poor households between 1996 and 2002. A further £325,000 was allocated via the council's Housing Investment Programme from 2000 to 2005 (Woking Borough Council, 2003). Aid was particularly directed to those who were spending 10 per cent or more of their household income on heating, the fuel poverty threshold as defined by the UK government (Greenpeace, 2006).

Lessons learnt

The major barrier to Woking's energy system was the regulations set by the Department of Trade and Industry (DTI) (Brown, 2004). The existing regulatory regime severely limited the number of domestic customers that could be supplied with green energy (Taking Stock, 2003). And although Woking's CHP and renewable energy generators were embedded in the local distribution network, they attracted similar trading charges as conventional central power stations because they were still connected to the national grid.

The use of the grid incurs transmission and distribution charges, the Fossil Fuel Levy, VAT (value added tax) and the Climate Change Levy for non-residential customers – adding to actual electricity charges and representing 25 per cent of most electricity bills. To avoid incurring the trading costs, Woking took advantage of the Electricity Orders of 1995 and 1997, which enabled them to lay their own private wires to all the properties that take its electricity, and to put into practice a local trading system. Combined with the efficiency of CHP, this made co-generation technology financially viable in Woking, enabling the supply of green energy to local customers at or below the market rate for brown energy (Woking Borough Council, 2003).

Some have argued that local decentralized energy production would mean reduced output. Woking has demonstrated that this is not so because the energy lost otherwise in transmission and distribution from remote locations is regained when electricity is generated locally. The localized 'private wire' district system can also operate in isolation when there is a disruption in national supplies.

Woking's plants mostly run on natural gas, tapped externally so this means that plants are still emitters of CO_2, albeit at lower levels than most fossil fuels (C40 Cities, 2008a). Emissions are compensated by the fact that when heat is recovered to either heat or cool buildings from local generation, it actually uses two-thirds less fuel and considerably less water (ABC Radio National, 2008). Woking is conscious of the brown fuel problem and has prompted new research into ways of generating their own renewable fuels (ABC Radio National, 2008).

Woking nevertheless enjoys many environmental and economic benefits from its energy system. Between 1991 and 2004, the town achieved a 48.6 per cent savings in energy consumption. CO_2 emissions fell by 77.4 per cent, NO_x emissions by 76.6 per cent and SO_2 levels by 90.9 per cent. Water consumption was also cut by 43.8 per cent, saving the council more than £5.4m on water bills over the period (Muir, 2005).

Woking cut out transmission and distribution charges discussed earlier by effectively cutting out the 'middle-men'. In undercutting the grid, green energy is supplied at 1 pence a unit cheaper than grid energy, hence increasing municipal revenue. Households enjoy reductions in energy bills by £130–200 a year and fuel-poor households are supplied with heat and power that cost a third less than before (Greenpeace, 2006).

The projects were quickly implemented due to support from private finance. By establishing an energy services company, it made it easier for people such as developers to design finance, accelerate, build and operate decentralized energy systems (ABC Radio National, 2008). Woking Council's partnership with an energy company made sure that locals did not share the high initial cost of the primary energy plant. This risk was transferred to Thameswey Energy Ltd, ensuring security in cost and supply. The company assumed responsibility not just for the design, finance and implementation of the plant but also the maintenance and production of green energy, which also included stand-by and top-up supplies (Woking Borough Council, 2003).

In recent years, Woking has founded an energy conservation and solar centre for future R&D projects (Morgan and Thorp, 2005) There are plans to increase the number of CHP stations within Woking and elsewhere in the UK (C40 Cities, 2008a). Woking Borough Council is also set to continue work with other cities, states and countries around the world in developing localized distribution networks (Stotz, 2005).

The Woking model demonstrates that renewable technologies are complementary and flexible – a type of engineering that can be assembled part-by-part as funds permit over time. It shows that renewables can be competitive but only when free from restrictions of centralized regulation and infrastructure (Greenpeace, 2006). However, most critical to the development of renewable energy was corporate commitment, cross-political party agreement and coordination, energy efficiency and robust support from a range of financial structures.

Reykjavik, Iceland

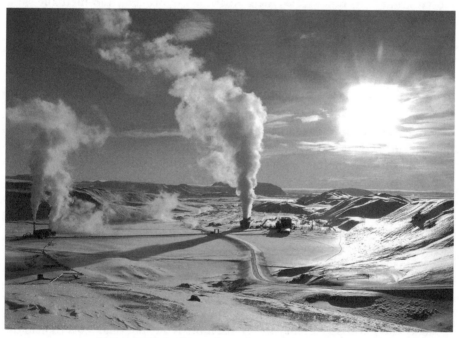

Figure 6.14 Krafla geothermal power plant in Iceland

Source: Ásgeir Eggertsson

Basic facts

Reykjavik is the capital city of Iceland, with a population of around 180,000 inhabitants, representing half the national population. As the location of the world's largest geothermal DH system, the entire city is powered by geothermal energy via an extensive electricity distribution network harnessing

750MW thermal power from steam, and a water distribution system generating 60 million cubic metres of hot water. The use of this natural resource has significantly reduced the city's reliance on fossil fuels (C40 Cities, 2008b).

Indeed, the city is located in a most advantageous position – a country with naturally occurring geothermal heat supplies due to vast landscapes of high volcanic activity. The natural heat from the underground hot water reservoirs is easily turned into clean heating and electricity for the entire country. In 2005, five geothermal power/heating plants were already producing 26 per cent of the country's electricity and 87 per cent of the housing and building heating needs (C40 Cities, 2008b). Furthermore, around 85 per cent of electricity in Iceland is generated by two hydropower stations generating a combined output of 11.4MW (Orkuveita Reykjavíkur, 2008).

It is unsurprising that Iceland is on track with plans for a hydrogen-based economy by 2050. To date it has installed a public transport system that mostly runs on renewable fuels such as hydrogen, bio-methane and compressed natural gas (CNG), fuels generated entirely by geothermal and hydropower.

How it evolved and what drove it

Since 1930, natural hot water has been used to heat buildings and homes in Reykjavik. In the late 1960s a new Energy Fund, which combined the former Electricity Fund and the Geothermal Fund, was established to further increase the use of geothermal resources in the country. Loans were granted to companies for geothermal exploration and drilling. These were transformed into grants if drilling failed to produce the expected results (C40 Cities, 2008b).

In 1990, the city built the Nesjavellir geothermal station 27km away to provide hot water for the Reykjavik area by heating freshwater with geothermal steam and hot water exchangers. In 1998, the power plant began operations with two 30MW steam turbines. In 2001, a third turbine was installed and the plant enlarged from a capacity of 90MW to 120MW in 2005; today total generation is 674GWh. The latest is a second, 90MW plant, built in 2006 at Hellisheidi (C40 Cities, 2008b).

The energy crisis of the 1970s prompted the national government to develop cheaper and cleaner energy and thus increase Iceland's security by reducing its dependence on fossil fuel imports (NATO, 2001). The crisis also triggered a new research project by the University of Iceland into hydrogen as transportation fuel. The plan was to transform all of the country's vehicle fleet (cars, trucks, buses) as well as fishing fleet to hydrogen internal combustion and hydrogen fuel cell power. Iceland's abundant sources of renewable energy and experience in exploiting geothermal resources made it the best location to develop a hydrogen technology not fed by fossil fuels (BRASS, 2008).

By 1998, the government had introduced a national hydrogen policy, supported by the university, which had developed and marketed the research gained in hydrogen production. Because there was no precedent for fuel cell technology using hydrogen or developing vehicles to run on hydrogen in Iceland, the country promoted itself abroad as the perfect 'testing ground' for this type of research, over time attracting interest from a variety of overseas companies. There were even plans with Japanese partners to develop a hydrogen shipping project (BRASS, 2008).

In 2003, the world's first hydrogen refuelling station was opened on the site of a conventional gas station in Reykjavík. From 2003 to 2008, four hydrogen buses operated as part of Reykjavik's public transport system, funded by the EU through the ECTOS/CUTE programme (BRASS, 2008). This allowed for several assessments in relation to the technical performance of buses, the effectiveness of hydrogen infrastructure in Iceland, and the socio-economic and environmental implications of using hydrogen as an energy source. By 2005, another hydrogen fuel station was in operation producing hydrogen on site through electrolysis. It is envisaged that by mid-2009, around 30 hydrogen-powered vehicles will be in use (Orkuveita Reykjavíkur, 2008).

In addition to hydrogen-based vehicles, there are 50 CNG vehicles, 46 bi-fuel vehicles and 4 dedicated heavy-duty natural gas vehicles in operation in Iceland. The bio-methane comes from Reykjavik's landfill site, operated by the waste management company SORPA (Orkuveita Reykjavíkur, 2008).

The UN University Geothermal Training Program (UNU-GTP) and Iceland's National Energy Authority established a training programme in 1968 to help countries with significant geothermal resources build specialists on geothermal exploration and sustainable development via specialized training. By 2005, 338 scientists and engineers from 39 countries had completed training. Today UNU-GTP graduates are among the leading specialists in geothermal R&D.

A research fund was set up by the local utility, Reykjavík Energy in collaboration with the universities in Iceland's capital for energy and environmental research. Each year Reykjavík Energy contributes approximately 0.5 per cent of its revenues towards the fund, as an addition to the initial capital contribution of ISK100 million ($783,700) (Orkuveita Reykjavíkur, 2008).

Lessons learnt

Although Iceland produces more electricity per capita than any other country in the world, less than 20 per cent of the country's energy potential is actually used. Iceland could produce as much as 50TWh of electricity, enough to provide heating and electricity services to 6 million consumers. Although technically feasible, the laying of underwater power cables to Scotland for example would not be economically justifiable. Electricity surpluses cannot be easily sold into the European grid, let alone to North American markets where distribution distances are too great (NATO, 2001). Iceland's government is conscious of this and is presently investigating ways to export this excess energy or to store it.

The people of Iceland have been weary of dam construction associated with the hydroelectric plants, lamenting the impacts it has had on ecology and landscape. However, hydropower experts have argued that hydroelectric power is cleaner than in other countries because of the cold climate of Iceland, where manmade lakes do not produce large amounts of algae that can cause severe environmental problems (NATO, 2001).

Some environmental groups have also opposed the construction of geothermal plants because of the greenhouse gases emitted by these plants. For example, about 7500 tons of CO_2 is released into the air each year from the Nesjavellir Power Plant (C40 Cities, 2008b). While geothermal electricity plants do emit some volcanic gases with greenhouse effects, these levels are much lower than a fossil fuel-driven power plant of similar capacity (NATO, 2001). It is estimated that since 1944, reductions in CO_2 emissions in Iceland have totalled 110 million tons. Between 2.5 and 4 million tons have been saved each year (C40 Cities, 2008b).

The reason for hydrogen as transport fuel was its clean-burning properties – water being the only by-product. While hydrogen is presently generated from geothermal and hydropower, the gas can alternatively be harnessed by directly drilling boreholes into the earth's crust and drawing it out from volcanically active zones. But this natural occurring gas has to be cleaned before it is used, which can be expensive. There are plans for drawing methanol gas from the exhaust gases from smelters, thus significantly reducing CO_2 emissions from those smelters, but Icelandic authorities see this as only an intermediate step (NATO, 2001).

The challenge for Iceland's hydrogen project is to make the replacement of petrol-driven cars with fuel cell-powered vehicles more feasible and cost effective. This will require a completely new system for distributing fuel, including fitting current petrol stations with the capacity to distribute charged batteries. The initiative will naturally demand significant public education and adaptation (NATO, 2001).

The skills exhibited by local companies during the early stages of implementation were an issue for most projects. The hydrogen vehicle project, for example, revealed that local stakeholders were lacking in technical expertise in fuel cell technology, but this was slowly overcome by forming partnerships with external companies. In general, projects were quickly and efficiently implemented thanks to the overall cohesive nature of Icelandic society. There were trusting relationships between the government and other stakeholders, and widespread public support for all renewable energy projects (BRASS, 2008).

Another challenge for projects was the time needed and the patience and motivation required from all parties, whether from universities, private companies, NGOs or government entities, to develop new innovative technology and to resolve any potential problems. For example, lengthy work was required on testing personal vehicles using the hydrogen technology before it was finally introduced to the wider public. It took time to train local people in the maintenance and repair of hydrogen fuel cell systems (BRASS, 2008).

Between the period of 1944 and 2006, financial savings were calculated to be around $4290 million, compared to the cost of heating by oil. Today the price of geothermal water is one third of the cost of heating with oil. The low cost of geothermal power has attracted a number of industries to Iceland, especially in the aluminium sector, where electricity is a significant addition to the production costs. In terms of evaluating total initial investments, this has been difficult to ascertain because operations began on a small-scale back in the 1930s. Larger developments occurred during World War II, in 1960 and between 1970 and 1980. It is estimated that to build the entire system today it would cost around $773 million (C40 Cities, 2008b).

Geothermal energy has led to Iceland's transformation from one of the poorest nations to one that enjoys a very high standard of living (C40 Cities, 2008b). The renewable energy industry has created many new jobs from energy production through to distribution and servicing. The Icelandic government was key to the success of renewable energy integration in Reykjavik, all of its actions demonstrating a steadfast commitment to the renewable energy industries. Local utilities as key energy providers have provided social and educational support. Reykjavik Energy has helped open the Museum of Energy as well as an educational centre. Its geothermal plants and other associated facilities attract a large number of visitors. It is estimated that 500,000 visitors travel to Reykjavik each year just to learn about the city's energy model. They vary from students, environmentalists and economists through to media, heads of state and other dignitaries (Reykjavik Energy, 2007).

The Icelandic model has shown how government commitment, perseverance, fearlessness and enterprise were key to the development of renewable energy in Reykjavik and the rest of Iceland. Community support and involvement of local and foreign companies in investing capital and technical expertise, have all contributed towards the creation of the most sophisticated geothermal facility in the world. The boldness of undergoing multiple energy transformations, from fossil fuels to geothermal, and then to hydrogen has resulted in an even cleaner, safer and healthier economy and environment.

Acknowledgements

I am grateful to Robyn Polan, University of New South Wales, for initial research support.

References

ABC Radio National (2008) 'Allan Jones: Getting off the grid', www.abc.net.au/rn/saturdayextra/stories/2008/2314663.htm

Acciona Energía (2007) 'Company Information', www.acciona-energia.com

Acciona Energía (2008) 'Photovoltaic solar', www.acciona-energia.com/default.asp?x=0002020402&lang=En

ADB (Asian Development Bank) (2003) 'Developing affordable renewable energy for remote Pacific Islands', www.adb.org/Media/Articles/2003/2677_Pacific_Affordable_Renewable_Energy/

Aicher, G. (2005) 'Navarra regional RES-e map: Electricity from renewable energy sources', http://209.85.173.104/search?q=cache:E2zmURIBtXAJ:www.res-regions.info/RESregions/fileadmin/res_e_regions/GN_RESe-map-NAVARRA-_EN__en.pdf+navarra,+spain+renewable+energy&hl=en&ct=clnk&cd=5

Aichernig, C., Hofbauer H., Rauch R., Bosch K. and Koch R. (2002) 'Biomass CHP Plant Güssing – A Success Story', http://members.aon.at/biomasse/strassbourg.pdf

AKI (ADN Kronos International) (2008) 'Energy: Italian mountain village finds the good line', www.adnkronos.com/AKI/English/Business/?id=1.0.1069653230

Alden, C. (2007) 'Creative energy', *Green Futures,* www.forumforthefuture.org/greenfutures/articles/creative-energy

Anjevall, J. (2007) 'Jämtland County', www.energiesystemederzukunft.at/edz_pdf/20070920_energieregionen_der_zukunft_folien_04_anjevall.pdf

AP Digital (2007) 'Spain's wind energy reaches new high', *The Age*, 21 March, www.theage.com.au/news/World/Spains-wind-energy-reaches-new-high/2007/03/21/1174153103988.html

Ashden Awards for Sustainable Energy (2008) 'Shaanxi Mothers, China', www.ashdenawards.org/winners/shaanxi

Beham, M. (2008) 'Case Study Multi-Mukli', www.arsenal.ac.at/downloads/CS/MultiMukli_engl.pdf

Binns, S., Osornio, J. P., Pourarkin, L., Pena, V., Roy, S., Smith, J., Smith, R., Wade, S., Wilson, S. and Wright, M. (2007) 'Power to the People', www.notre-europe.eu/uploads/tx_publication/Etud59-Columbia-PowertothePeople-en.pdf

Bioenergiedorf (2005) 'Bioenergiedorf, 2005', www.bioenergiedorf.de/con/cms/6/home/

BMVIT (2007) 'Model Region Güssing', www.eee-info.net/Download/20070430_Forschungsforum_GuessingEEE.pdf

BRASS (2008) 'Sustainable Energy Case Study: Iceland's Hydrogen Experiment', www.brass.cf.ac.uk/uploads/Sus_Community/Case_Study_Hydrogen.pdf

Brown, P. (2004) 'Woking Shines in Providing Renewable Energy', *The Guardian*, 26 January

Burgermeister, J. (2007) 'Renewable Energy Powers Italian Town and Its Economy', www.renewableenergyworld.com/rea/news/story?id=50863

C40 Cities (2008a) 'Energy: Woking, United Kingdom', www.c40cities.org/bestpractices/energy/woking_efficiency.jsp

C40 Cities (2008b) 'Renewables: Reykjavik, Iceland', www.c40cities.org/bestpractices/renewables/reykjavik_geothermal.jsp

City of Helsinki Ministry of the Environment (2005) 'Eco-Viikki: Aims, Implementation and Results', City of Helsinki, Helsinki

CunCaoXin (2005) 'A Green Revolution in Red China is Taking Root on Yellow Earth', www.cuncaoxin.org/y-x7.htm

CunCaoXin (2007) 'The beneficiaries of "Sunflower Action"', www.cuncaoxin.org/y-x2007-B.htm

CunCaoXin (2008a) 'Shaanxi Village Biogas Carbon Offset Calculation', www.cuncaoxin.org/y-x5.htm

CunCaoXin (2008b) '2008 Sunflower Action Prospering Industrial Plan', www.cuncaoxin.org/y-x2008-1.htm

Deutsches Generalkonsulat Melbourne (2005) 'Jühnde is Germany's First Bio Energy Village', www.melbourne.diplo.de/Vertretung/melbourne/en/05/Research__and__Technology/Energy__J_C3_BChnde__Seite.html

Douthwaite, R. (2006) 'From recession to renewables', *Construct Ireland*, vol 17, pp19–27

Du Bois, D. (2007) 'Spain's Navarre region runs with renewable energy', *Energy Priorities*, June, http://energypriorities.com/entries/2007/06/spain_navarre_renewable.php

ECO-City (2009) 'Tudela community, Spain', www.ecocity-project.eu/TheProjectTudela.

EFFP (2004) 'Successful COLLABORATION in practice', www.effp.com/documents/publications/Ten_International.pdf

EEE (European Centre for Renewable Energy Güssing) (2009) 'Solarschool Güssing', www.eee-info.net/Seiten_e/druck.php?text=100&titel=Solarschool%20G%FCssing

Energie-Cités (2000) 'Bioclimatism: Helsinki (Finland)', www.agores.org/Publications/CityRES/English/Helsinki-FI-english.pdf

Eurosolar (2004) 'Appreciation Municipality of Güssing', www.eurosolar.de/en/index.php?Itemid=27&id=210&option=com_content&task=view

Eurosolar (2005) 'Appreciation Municipality of Jühnde', www.eurosolar.de/en/index.php?Itemid=26&id=199&option=com_content&task=view

Eurosolar (2006) 'Appreciation Jämtland County Energy Agency',www.eurosolar.de/en/index.php?option=com_content&task=view&id=186&Itemid=25

Eurosolar (2007) 'Appreciation Acciona Solar S.A.', www.eurosolar.de/en/index.php?option=com_content&task=view&id=284&Itemid=1

Eurosolar (2008) 'Appreciation Thisted Kommune', www.eurosolar.de/en/index.php?option=com_content&task=view&id=285&Itemid=1

Fairless, D. (2007) 'Energy-go-round', *Nature*, no 447, pp1046–1048

Fangmeier, E, (2008) 'Local 100 per cent RE Supply: The example of the Bioenergy Village Jühnde Germany', www.gaccny.com/fileadmin/user_upload/Dokumente/Career_Services/Renewables_Presentations/Eckhard_Fangmeier__Read-Only_.pdf

Federal Ministry of Economics and Technology (2008) 'The first Bioenergy Village in Germany', www.german-renewable-energy.com/Renewables/Navigation/Englisch/Biomasse/case-studies,did=132906.html

Fitzgerald, J. (2008) 'Austrian city as local energy model', http://postcarboncities.net/node/3214

FOE (Friends of the Earth Hong Kong) (2008) 'Shaanxi Province Village Biogas Demonstration Project: Women power to green power', www.foe.org.hk/welcome/geten.asp?language=en&id_path=1,%2011,%203203,%203205

Gobierno de Navarre (2007) 'Navarre Renewables 2010 Horizon Energy Plan', www.biofuel2g.com/Argazkiak/Navarra_Renewables.pdf

Green Labels Purchase (2006) 'Good practice example', www.greenlabelspurchase.net/Varese-Ligure.html

Greenav (2008) 'Understanding Wood Pellet Boilers – A simple guide', http://greennav.wordpress.com/2008/04/30/understanding-wood-pellet-boilers-a-simple-guide/

Greenpeace (2006) 'Decentralising energy – the Woking case study', www.greenpeace.org.uk/files/pdfs/migrated/MultimediaFiles/Live/FullReport/7468.pdf

GreenUpAndGo (2007) 'Güssing: Renewable energy rejuvenates the town', www.greenupandgo.com/renewable-energy/gussing-renewable-energy-rejuventates-the-town/

Hagdahl, P. (2003) 'Invest in Jämtland, Why Jämtland?', www.midscand.com/filer/Invest%20in%20Jamtland.ppt

Hall, K. (2003) 'Comfortable use of wood pellets in one-family houses in Jämtland County', www.managenergy.net/download/nr6.pdf

Hall, K. (2004) 'Thermal solar energy in detached houses', www.managenergy.net/download/nr108.pdf

Hametner, M. (2006) 'Bio-energy Mureck', www.nachhaltigkeit.at/bibliothek/tatenbank/en/f0001066.pdf

Heiskanen, E., Hodson, M., Mourik, R. M., Raven, R. P. J. M., Feenstra, C. F. J., Alcantud, A., Brohmann, B., Daniels, A., Di Fiore, M., Farkas, B., Fritsche, U., Fucsko, J., Hünecke, K., Jolivet, E., Maack, M., Matschoss, K., Oniszk-Poplawska, A., Poti, B., Prasad, G., Schaefer, B. and Willemse, R.

(2007) 'Factors influencing the societal acceptance of new energy technologies', www.esteem-tool.eu/fileadmin/esteem-tool/docs/Resourcesreport.pdf

Ichaso, M. (2000) 'Wind Power Development in Spain, the Model of Navarra', *DEWI Magazine*, no 17, pp49–54

ICEPAC (2000) 'Bilateral and Multilateral Partnerships', www.ice-pac.org/background/Bilateral%20and%20Multilateral%20Donors.html

IDAE (2001a) 'EL HIERRO Island, biosphere reserve, 100 per cent RES supply', http://ec.europa.eu/energy/idae_site/deploy/prj042/prj042_2.html

IDAE (2001b) 'Varese Ligure 100% Sustainable', http://ec.europa.eu/energy/idae_site/deploy/prj083/prj083_1.html

INSULA (2001) 'Renewable Energy Islands: The Danish Energy Way', http://insula.org/islandsonline/REI%20-%20The%20Danish%20Way-1.pdf

INSULA (2008) 'El Hierro 100 % RES', www.insula.org/index.php?option=com_content&task=view&id=19&Itemid=33

Intelligent Energy Europe (2008) 'European sustainable urban development projects: Viiki', www.secureproject.org/download/18.360a0d56117c51a2d30800078421/Viikki_Finland.pdf

Iris Europe (2007) 'Iris Europe – Canary Islands', www.iris-europe.eu/IMG/pdf/IRIS_MT_CAN_cs17_soho-solo-Set07_1_.pdf

Kolbert, E. (2008) 'The Island in the Wind: A Danish community's victory over carbon emissions', *The New Yorker*, July

Korneffel, P. (2005) 'Fog's their only foe', *New Energy*, vol 6, pp34–39

Liebenthal, A., Mathur, S. and Wade, H. (1994) 'Lessons from the Pacific island experience', World Bank Technical Paper no 244, Energy series, World Bank, Washington, DC

Lungescu, O. (2007) 'Navarra embraces green energy', http://news.bbc.co.uk/2/hi/europe/6430801.stm

Lutter, E. (2004) 'Case Study: 2 MWel biomass gasification plant in Güssing (Austria)', www.ecd.dk/download/wp3/güssingaustria.pdf

Marconnet, M. (2007) 'Integrating Renewable Energy in Pacific Island Countries', http://researcharchive.vuw.ac.nz/handle/10063/491

McCormick, K. and Kåberger, T. (2007) 'Key barriers for bioenergy in Europe: Economic conditions, know-how and institutional capacity, and supply chain co-ordination', *Biomass and Bioenergy*, vol 31, pp443–452

Mitra, I. (2006) 'A Renewable Island Life', *reFOCUS*, November/December, pp38–41

Morgan, R. (2006) 'Renewable Energy Policy in Woking and the South East', www.spongenet.org/library/Woking%20BC%20Renewables%20Policy%20Oct06.ppt

Morgan, R. and Thorp, J. (2005) 'Implementing New Approaches to Energy in Woking', www.sustainablebuild.org/downloads/ray_morgan.pdf

Muir, H. (2005) 'Wake-up call from Woking', www.guardian.co.uk/society/2005/jun/29/environment.interviews/print

NATO (2001) 'Trip Reports: Visit to Iceland', www.nato-pa.int/archivedpub/trip/au280gen-iceland.asp

ökoEnergieland (2008) 'Historical Güssing', www.oekoenergieland.at/english-information.html?start=1

Orkuveita Reykjavíkur (2008) 'Reykjavik Energy and Environment', www.or.is/English/EnergyandEnvironment/

Padrón, T. (2004) 'La primerea isla del mundo 100% energies renovable', www.unescocan.org/100reshierroeng.htm

Piernavieja, G. et al[Q57] (2003) 'El Hierro: 100% res: An innovative project for islands' energy self-sufficiency', http://insula.org/islandsonline/ElHierroeng.pdf

Procura (2008) 'Case study: Varese Ligure towards 100 % Renewable', www.procuraplus.org/fileadmin/template/projects/procuraplus/files/CD-ROM/Case_Studies/Electricity_Varese_Ligure_Italy.pdf

Rauch, R. (2005) 'Energy Supply Concepts for the Region Guessing, Austria', www.aer.eu/fileadmin/user_upload/Commissions/RegionalPolicies/EventsAndMeetings/2005/Presentations_Norrbotten/R-Rauch.ppt

RENET (Renewable Energy Network Austria) (2006) 'Güssing', www.renet.at/english/sites/guessing.php

Reykjavík Energy (2007) 'A Dynamic Company – a Leading Power', www.or.is/media/PDF/ORK per cent2038077 per cent20Adalbaeklingur_ENS_Lowres.pdf

Reykjavik Energy (2008) 'Energy and Environment', www.or.is/English/EnergyandEnvironment/

RISI (2008) 'SCA joins forces with Statkraft to invest in wind power', www.risiinfo.com/magazines/pulp-paper/news/SCA-Statkraft-i...er%2Fnews%2FSCA-Statkraft-invest-wind-power-renewable-energy.html

Ruwisch, V. and Sauer, B. (2007) 'Bioenergy Village Jühnde: Experiences in rural self-sufficiency', www.bioenergiedorf.info/pdfs/Bioenergy%20Village%20(20-09-07).pdf

Salazar, A. (2008) 'WP2: Regional summary report prepared by Government of Navarra', www.efficient-electricity.info/efficient-electricity/fileadmin/efficient_electricity/WP_2/GN_WP2_Summary.pdf

Samsø Energiakademi (2008) 'Samsø – a Renewable Energy Island', www.energiakademiet.dk/images/imageupload/File/UK/RE-island/10year_energyrepport_UK_SUMMARY.pdf

SEEG (2004) 'Bioenergy Cycle', www.seeg.at/en/index.php

SEEG (2007) 'Bioenergy Murck', www.seeg.at/en/index.php

Siehr, M. (2007) 'BioProm-publishable result-oriented Report', www.bioprom.net/pdf/Berichte/EN/Publishable_result_oriented_report_Aug_2007

SOLPROS (2003) 'Ekoviikki Sustainable City Projects', www.kolumbus.fi/solpros/ekoviikki.htm

Stotz, E. (2005) 'Woking: Municipal Government', www.iclei-europe.org/fileadmin/user_upload/ClimateAir/Woking.doc

Sustainable Energy Europe (2008) 'Changing mind, changing behaviour: Swedish county converts from oil to biomass, from fossil fuels to renewables', www.sustenergy.org/UserFiles/File/6_Jamtland_PD(1).pdf

Taking Stock (2003) 'Case Study 2: Woking Borough Council Energy Services', www.climatespace.org/wp-content/uploads/2007/10/case_study_2-woking.pdf

Technology Review (2007) 'Solar Energy in Spain', www.technologyreview.com/microsites/spain/solar/

Thisted Municipality (2008a) 'Thisted Municipality: clean energy – clean nature', http://climate.thisted.dk/wp-content/uploads/thisted-municipality-climate-friendly-region.ppt

Thisted Municipality (2008b) 'Thisted Municipality', http://climate.thisted.dk/

Thisted Municipality (2008c) 'The Thisted Model', http://climate.thisted.dk/2008/09/the-thisted-model/

Thisted Municipality (2008d) 'District Heating from Black Diamond', http://climate.thisted.dk/2008/10/district-heating-from-black-diamond/

Thisted Municipality (2008e) 'Town Meeting in Thisted', http://climate.thisted.dk/2008/09/town-meeting-in-thisted/

Thisted Municipality (2008f) 'The Thisted Movie' (Quicktime movie), http://climate.thisted.dk/gb/downloads/

Thisted Municipality (2008g) 'First Danish Style Passive House', http://climate.thisted.dk/gb/2008/09/first-danish-style-passive-house/

Thomas, J. (2008) 'Italian town runs on 100% renewable power', www.metaefficient.com/renewable-power/italian-town-runs-on-100-renewable-power.html

Tirone, J. (2007) '"Dead-end" Austrian town blossoms with green energy', *International Herald Tribune*, 28 August

TV LINK Europe (2005) 'Jämtland County – A Region Fuelled By Biomass', www.tvlink.org/vnr.cfm?vidID=159

Tyskling, K. (2008) 'Biobränsle från det jämtländska jordbruket', Dept. of Economics, SLU. Examensarbete / SLU, Institutionen för ekonomi vol 513

UNESCAP (2002) 'Conclusions and recommendations from the Workshop on Sustainable Energy Policies and Strategies for Pacific Island Developing States, Suva, Fiji, 4 to 5 February 2002', www.unescap.org/esd/energy/cap_building/renewable/documents/Conclusions%20and%20recommendations.pdf

UNESCO (2003) 'Rays of hope: renewable energy in the Pacific Islands', http://portal.unesco.org/en/ev.php-URL_ID=43115&URL_DO=DO_TOPIC&URL_SECTION=201.html

United Nations 'RENEWABLE ENERGY TRAINING in PACIFIC ISLAND DEVELOPING STATES', November 2004. Retrieved 14 October 2008 from http://www.unescap.org/esd/energy/cap_building/renewable/documents/Renewable_Energy_Training.pdf

Urien, B. (2007) 'Wind Energy in Navarre: current situation and future challenges', www.europeanislands.net/docs/Navarra.pdf

Urschick, A. (2004) 'European Center for Renewable Energy Güssing Ltd (EEE)', www.europeangreencities.com/pdf/activities/ConfNov2004/summaries/10_Energy%20innovations%20in%20Gussing%20as%20economic%20engine.pdf

Walsh, B. (2008) 'Heroes of the Environment 2008: Soren Hermansen', *Time Magazine*, October

Wheldon, A. (2006) 'Fuel, compost and sanitation from biogas in rural China', www.ashdenawards.org/files/reports/Shaanxi%202006%20Technical%20report.pdf

White, E. (2007) 'Bio-fuel Revolution Part 2', www.youtube.com/watch?v=LKyhrZwjCD8

Wikipedia (2008a) 'Güssing', http://en.wikipedia.org/wiki/Güssing

Wikipedia (2008b) 'Jämtland', http://en.wikipedia.org/wiki/Jämtland

Wikipedia (2008c) 'Crop rotation', http://en.wikipedia.org/wiki/Crop_rotation

Woking Borough Council (2003) 'Sustainable Woking: Background case studies', www.woking.gov.uk/environment/climate/Greeninitiatives/sustainablewoking

Woodruff, A. (2007) 'An economic assessment of renewable energy options for rural electrification in Pacific Island countries', www.sopac.org/Rural+Renewable+Energy+Economics

WSA (Welsh School of Architecture) (2004) 'Case study: Viikki eco neighbourhood blocks', www.cardiff.ac.uk/archi/programmes/cost8/case/holistic/viikki.html

Chapter Seven

Feed-in Tariffs: The Policy Path to 100%

Miguel Mendonça and David Jacobs

The renewable energy FiT mechanism has emerged as the most effective, affordable and flexible means to introduce renewable power quickly. Nations, regions and even large cities can use its principles with confidence to deploy any technology or technologies they wish to in their jurisdiction, in order to more effectively harvest the renewable resources therein. Caps and time limits on the programme can be set to achieve certain goals on deployment and overall programme cost, but leaving this open will allow more stable and vigorous investment and job creation to occur.

To reach a 100 per cent supply of energy from renewables, a FiT support scheme can be used to boost market development rapidly. If a country or region is serious about moving to renewable energy, it can set aggressive targets and payment rates for the different technologies most suitable for the region. World markets for renewables are still in the establishment phase, so FiTs can help nations to acquire global market share, as well as drive domestic green industries. Importantly, FiTs ensure that costs are controlled through the technological development phase, and they only reward efficient, fully-functioning installations.

How do FiTs work?

In the electricity sector, FiTs are an efficient instrument, with an ideal mix of low cost, simplicity and effectiveness. They provide a legally guaranteed long-term payment to producers of renewable energy, for each kilowatt-hour of energy they feed into the grid. Payment levels are set according to the production costs for each technology, and usually decline by a set percentage each year – the so-called tariff degression – in order to anticipate technological development. For example, you would receive x cents/kWh for 20 years, but if you build next year, you receive x -5 per cent.

Costs are usually covered by a small increase in the electricity bills of all customers in the nation or state. The legally guaranteed production payment for each technology ensures investment stability, and makes financing cheaper and easier to acquire. The Stern Review found FiTs to be superior to other support schemes in terms of cost and deployment (Stern, 2006), and the European Commission's 2005 study showed they are a better market launch mechanism (European Commission, 2005). Indeed, many empirical comparative studies have found the same (Mendonça, 2007).

Germany has demonstrated the dramatic potential of FiTs for spurring deployment, job creation, international exports and carbon savings (BMU, 2008). The German FiT law, the Renewable Energy Act (known as the EEG, or Erneuerbare-Energien-Gesetz) has offered investment security for almost two decades. Introduced for the first time in 1990, it has been constantly adjusted and improved. More and more renewable energy technologies have been taken into consideration and thus the share of renewable electricity has been considerably increased.

Spain and formerly Denmark have also used the instrument to build major global market share in technology exports. FiTs provide a successful mechanism to create a national renewable energy industry. Six of the ten largest wind turbine manufacturers are from Germany, Spain or Denmark (WWEA, 2007). Other nations in this top-ten list, especially China and India, also implemented national or regional price-based support schemes in order to create a strong national industry (Lewis, 2007).

As the share of renewable energies in the energy mix increases, it becomes increasingly important to implement design options that facilitate the integration of 'green' electricity in the 'grey' power market. Therefore, several options can be considered to tackle the issue of variability of some technologies – wind and photovoltaics in particular – and to better match supply and demand. For the latter, so-called premium-FiTs have been implemented in several European countries. Here, the renewable electricity is sold on the spot market, and in addition to the market price the power producers receives a fixed premium-FiT payment that covers the cost difference between the market price and the actual generation cost for each renewable technology. Usually premium-FiTs are introduced as an optional alternative to the fixed tariff payment. Generally, only relatively large players opt for the premium-FiT option, since the effort for selling electricity on the spot market will be disproportional for small producers, such as private households with a photovoltaic installation on the roof.

FiTs in the developing world

In several developing countries FiTs have been introduced but there are clear issues of affordability to be addressed for the poor to not be burdened by any additional payments required to secure renewables-based electricity supplies. Alternatively, price fluctuations in the fossil fuel markets have already added great pressure on the consumers of kerosene for cooking, lighting and heating – and in the medium term renewable supplies could offer a relief by stabilizing the source costs. If needed, the extra cost could be distributed equally among all electricity consumers, as in most European FiTs, but since price increases are often politically difficult to introduce in normally highly subsidized markets, a national fund – partly financed by international donors – will often have to be set up. It is also possible to combine both fee increases and fund support, as Kenya has done. The Kenyan FiT focuses on relatively cost-efficient technologies, including biomass, wind power and hydropower. Mauritius introduced a FiT for the sugar industry, which produces electricity from bagasse, a by-product of the sugar production process. This way, more than 40 per cent of the country's electricity demand is covered. Today, sugar cane producers make more money with electricity generation than with the actual sugar cane production.

FiTs have great potential for tackling domestic fuel poverty by providing independence from energy bills to the owner. Combined with energy efficiency improvements, the generation and export of energy can act as an effective hedge against future energy price increases. The technologies themselves could be paid for through deeming and capitalization of the FiT, complemented by soft loans from government or other agencies.

Transport and heat sectors

The transport sector can see rapid electrification of vehicles if a combination of measures is pursued. A FiT can be set for new technologies, such as under-road magnetic generation (MagKinetics, 2008), small roadside maglev wind turbines (Xinhua News Agency, 2007) or weight from traffic (Kim et al, 2008). Thermal capture technology can be deployed to use solar heat from the surface of roads to generate electricity (ENN, 2008). Research, development and demonstration of the technologies themselves can be supported through government funds or partnerships with investors and developers. Since FiTs only pay for actual energy generated, the technology has to work effectively, and there is no risk to taxpayers who fund the FiT.

With regard to renewable heat, at the time of writing, only Germany had so far introduced legislation, but opted for an obligation mechanism instead that makes use of an existing grant or loan programme, citing cost, complexity and constitutional issues with a FiT. Other countries, including the UK and France, are beginning to explore the FiT route to renewable heat market development.

Closing thoughts

Currently some 50 countries, states and provinces around the world are turning to FiTs, or returning to them, to get their nation moving forward, faster, towards a higher penetration of renewables (REN21, 2007). Although many barriers stand in the way, and many scenarios are still putting potential 100 per cent targets at several decades away, it can be done (Boyle, 2008).

Germany's growth has put it ahead of the 5.2 per cent per annum growth rate that the United Nations International Sustainable Energy Association (UNISEO) has reported (2008) will be necessary to replace coal and nuclear by 2050, while accounting for increasing global energy demand.

FiTs, combined with adequate supporting conditions, including planning, approvals and grid connection, can generate a great deal of this growth, and not only in industrialized countries with the domestic capital to fund the growth. Developing countries, too, can introduce this powerful tool by working with donors and lenders, community development groups and establishing policies permitting a gradual reduction or transfer of fossil fuel subsidies to match progressive renewable energy development policies. Cost competitiveness with the subsidized conventional energy industry is more likely to result from a widespread and aggressive global uptake of renewables, and the stable long-term appeal of FiTs is suitable for this task.

References

BMU (German Federal Environment Ministry) (2008) 'Renewable energy sources in figures: National and international development', www.bmu.de/files/english/renewable_energy/downloads/application/pdf/broschuere_ee_zahlen_en.pdf

Boyle, G. (2008) 'Positive outlook', *Energy Engineering*, vol 18, pp13–14

ENN (2008) 'Solar collector could change asphalt roads into renewable energy source', www.enn.com/sci-tech/article/37929

European Commission (2005) 'Communication from the Commission: The Support of Electricity from Renewable Energy Sources {SEC(2005) 1571}', http://ec.europa.eu/energy/res/biomass_action_plan/doc/2005_12_07_comm_biomass_electricity_en.pdf

Kim, R. H., Kim, B., Kim, M. S. and Kim, D. Y. (2008) 'Power generation pad using wasted energy', www.freepatentsonline.com/7432607.html

Lewis, J. (2007) 'A comparison of wind power industry development strategies in Spain, India and China', prepared for the Center for Resource Solution, San Francisco

MagKinetics (2008) 'Technology', www.magkinetics.com/technology.html

Mendonça, M. (2007) *Feed-in Tariffs: Accelerating the Deployment of Renewable Energy*, Earthscan, London

Mendonça, M., Lacey, S. and Hvelplund, F. (in press) 'Stability, participation and transparency in renewable energy policy: Lessons from Denmark and the United States', *Policy and Society*, in press, http://dx.doi.org/10.1016/j.polsoc.2009.01.007

REN21 (Renewable Energy Policy Network for the 21st Century) (2007) 'Renewables 2007 Global Status Report', www.ren21.net/pdf/RE2007_Global_Status_Report.pdf

Stern, N. (2006) *Stern Review: Report on the Economics of Climate Change*, Cambridge University Press, Cambridge

UNISEO (United Nations International Sustainable Energy Association) (2008) 'Blueprint for the Clean, Sustainable Energy Age', www.uniseo.org/blueprint.html

WWEA (World Wind Energy Association) (2007) *Wind Energy International 2007/ 2007*, Wind Energy International, Bonn

Xinhua News Agency (2007) 'China to mass produce maglev wind power generators', http://news.xinhuanet.com/english/2007-11/05/content_7016626.htm

How to Achieve Renewable Energy Regions and Advance Sustainable Development: Integrated Models and Processes in Germany

Peter Moser, Lioba Kucharczak
and Cord Hoppenbrock

The finite nature of fossil and nuclear energy sources is becoming increasingly apparent. In the medium and long term there is no alternative to the complete reliance on renewable energy sources. The feasibility of this conversion is no longer questioned, but actually requested by a growing number of scientists and politicians. This is not a question of 'whether' but 'when' it will happen. In this regard, regions have a decisive function to stimulate a comprehensive changeover to the solar age. Social and economic players, government guidance, different renewable energies technologies, economic dynamics as well as strategic development perspectives play an important role. This chapter presents a regional overall model, a guideline on how to achieve 100 per cent renewable energy regions and regional examples from Germany.

100 per cent renewable energy regions and the path required to achieve this status are the subject of an ongoing research project funded by the Federal Ministry for the Environment, Nature Conservation and Nuclear Safety. This project is conducted by the Competence Network for Decentralized Energy Technologies (deENet GmbH) in cooperation with the University of Kassel. Besides scientific and theoretical aims, the objective of the project is to support and evaluate regional activities (DeENet, 2009).

Regions as space for players and the implementations of 100% renewable energies

To date mainly the federal and the state levels have played a significant role in boosting the use of renewable energies in Germany. Currently, all regional activities depend highly on national frameworks that enable the profitable use of renewable energies. Such frameworks – for example, EEG in Germany – are not explored in any depth here.

At the regional and local levels people have been content with isolated activities of limited impact but high publicity. But meanwhile a growing number of players recognize the unique role of cities and regions in climate protection and alternative energy supply. The 'region' is a complex geographic territory of multi-dimensional interrelated structures. For a multitude of reasons, regions are the appropriate level for an extensive increase of a renewable energy supply. Along with the regional scale, the local level element will play a decisive role in developing climate strategies integral to regional planning and implementation.

The prestigious climate researchers Rahmstorf and Schellnhuber (2007) also support the thesis that the urban system is the ideal geographic entity to organize integrated solutions for the climate problem. Here appropriate prevention and adjustment measures can be planned, trialled and developed in direct dialogue with important players. The scientists argue that small-scale units are on the one hand transparent enough to allow efficient implementation, but on the other hand they are large enough to transform individual motives and activities into directed and powerful cooperation processes. Their conclusion, 'Medium is beautiful!', approves of the regional level as the decisive space for players and implementation – here the change to 100 per cent renewable energies promises to be feasible soon. In particular, structurally weak, low-density regions find a regional development driver in the increase of renewable energies – and along with this they find hope to buffer future energy price risks.

Regions provide enormous scope for strategic concepts and creations within evolving legal frameworks and the innovative application of financial resources. Regional decision-makers often have a strong local groundedness and hence like the vision of 'energy independence' or '100 per cent renewable energy regions' to develop a strong regional sense of identity (Tischer et al, 2006). They can be key champions of renewable energy proliferation. Visible changes, socially and spatially integrated, support these processes. Renewable energy expansion on the regional planning level invites the rallying of a number of special interest and lobbies behind this cause, including those of landscape protection, nature conservation and ecological sustainability. By engaging the population through associations in which the citizens become partners and customers of regional energy supply companies – the interests of different economic players can be integrated as well.

Economic factors provide further arguments for a comprehensive expansion of decentralized renewable energies. Against the background of the existing economic system, the so-called free-market economy, decentralized energy systems have to be seen in terms of economic laws, and therefore they require implementation in an optimal scale. Undersized, i.e. small-scale or isolated designs incur the opportunity costs of not benefiting from economies of scale, while the over-dimensioned, supra-regional grids and networks incur excessive transport costs and lower the system's efficiency. This applies especially, but not only, to the heating and cooling sector.

The application of a range of technologies in a regenerative energy mix involves different players as well as spatial perspectives. The smallest spatial level is the individual building – this field of activity already offers tremendous opportunities for the utilization of renewable energies. The next largest scale is that of the quarter, village or district. Mainly DH (or, in tropical areas, cooling) and, to a lesser extent, co-generation networks are found here. A 100 per cent supply within the electricity sector, however, is usually only feasible on the regional level and the introduction of electric drive vehicles, which run on renewable energies, will probably need an even larger scale of operation, since vehicles also move out of their region of origin (see Figure 8.1).

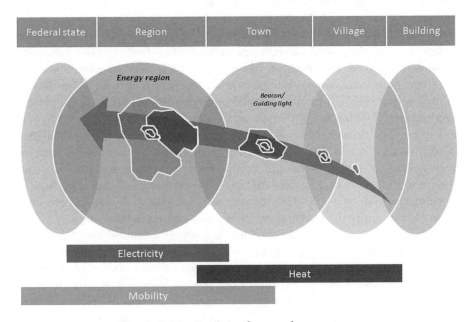

Figure 8.1 Regional significance of energy types

Source: Moser and Hoppenbrock (2008)

These examples illustrate the required openness of a regional system, which on the one hand has to provide a decentralized energy system for the region in question, and on the other hand has to allow imports of electricity from other regions (for example, from offshore windparks) – at least temporarily – to achieve the 100 per cent renewable supply. In terms of regional boundaries, the phrase 'region' should hence be treated very flexibly. In addition to the existing spatial-organizational delineation of regions by existing programmes, political expectations and planning controls, also cultural and society-oriented aspects play an important role for energy regions.

The objective of a full supply – a 100 per cent target – suggests a regional approach, as only at this scale can profitability for the required number of generation points be expected. At the outset, smaller-scale units such as quarters or municipalities can be of interest as well, since these can act as 'breeding grounds' or 'beacons' for a larger scale development process (see Figure 8.1). But these approaches tend not to be economical – they need the integration into larger contexts with a regional energy provider.

100% renewable energies in electricity, heat and mobility

A full regional supply with renewable energies is mainly obtainable through the combined application of various renewable energies and drastically reduced energy consumption, which also can be achieved by raising efficiency and saving energy. The foundation for an increase of renewable energies generation capacity and the proliferation of renewable technology in a region is the development of a sensible, decentralized, renewable energy system. Unlike the present, single-directional system of a few massive generators networked to feed a myriad of passive users, the new paradigm is about fostering numerous single but activating decisions: private photovoltaic rooftop systems; biogas plants operated jointly by several farmers; cooperative wind farms, in which many players of a region can participate as investors – and so on. The decision-making powers and timeframes for the establishment of plants can and will differ. They range from simple applications, which can be prepared in a few weeks, to comprehensive permit procedures, which require several years of development time. The most important precondition to reaching the goal of a 100 per cent renewable region is the determination to summon and apply the regional potentials in a comprehensive manner.

A growing number of German municipalities, counties and regions are meeting these challenges, resolving to develop a 100 per cent renewable energy target for their regions. Their 100 per cent targets at first often concentrate on the electricity sector, but at the end of the process it can often cover heat and mobility aims as well. Other important targets for regions are to reach 'zero-emission' or 'climate neutrality'. These objectives include a 100 per cent renewable energy target as part of an even more comprehensive climate protection target (for example, the Zero-Emission-Village Weilerbach).

Electricity

Germany leads in Europe when it comes to increasing wind and photovoltaic energy supply. Renewable sources provide more than 14 per cent of German electricity. But for some time now the national renewable expansion objectives no longer satisfy many cities and regions; they aim at achieving a full supply with renewable electricity, and rapidly.

Recently the long-known, theoretical potential for a full supply with different renewable systems in the German electricity sector has been tested in an industry pilot dubbed the 'Combined Renewable Energy Power Plant'. The 'Combi-Plant' demonstrated that a purely renewable, distributed electricity supply responding to real-time demand is technically and economically feasible on a decentralized level and can be ensured to 100 per cent. The Combi-Plant is based on digital networking and a centrally controlled balancing of different decentralized renewable energy power plants. Eleven wind turbines, 4 biogas plants, 20 photovoltaic systems and one pumped-storage hydropower plant were connected with each other via a central control unit. Thereby it was demonstrated on a small scale – a ten-thousandth of the German electricity demand – that purely renewable energy production can be adapted to the actual energy demand in 15-minute intervals. It was shown in this way that wind, water, sun and biomass complement one another powerfully (www.kombikraftwerk.de).

Heat

The full supply with heat energy will be achieved on a rather small scale and based almost entirely on bioenergy. So-called bioenergy villages, deriving their biomass heat sources from their surroundings, work extremely well in Germany (for example the bioenergy villages of Iden, Oberrospe, Rai-Breitenbach or Mauenheim). But in the planning and construction of the plants they often did not pay attention to any efficiency criteria, such as thermal insulation measures. Therefore the criteria of sustainability have to be considered more carefully before the future construction of energy plants (Moser, 2008a, 2008b; Scheffer, 2008).

Mobility

Renewable mobility is currently largely available only via bioenergy, such as plant oil, biogas or ethanol. But the energy and climate efficiency of these bioenergy products is poor, as vehicle waste heat remains unused. Great opportunities are offered in the shift to electric vehicles at the regional level. In addition to affording emission-free mobility they offer great opportunities in energy storage, load management and dispatch, as in the shaving of peak demands or the balancing of energy supply.

Mission statement 'ideal region': What are the most important milestones?

Regions that wish to supply themselves entirely through renewably sources need to embark on clear and determined strategies consisting of goals and the steps required to achieve them. The integrated model of an ideal region is therefore presented, designed as a desirable future vision and mission statement, using diagrammatic scenarios.

Mission statements express strategic future objectives, to help coordinate action. They are used in business, politics, science and increasingly also in planning, where they provide agreed-upon conceptual frameworks for decision-makers and the public. Scenarios have recently emerged as effective tools of planning, and they are frequently used in the process of developing mission statements (Moser and Meyer, 2002). They are sketchy but useful in integrating various aspects of the future in tangible ways, as possible outcomes of hypothetical course of actions or events. Scenarios serve to make potential futures understood, both those that continue a present trend, and those that envision alternative futures. They can provide visions for a comprehensive development of entirely renewable regions – assuming that these scenarios are both desirable and feasible. Forecasting scenarios can be used as well as those that attempt the back-casting of steps or milestones required to lead to certain outcomes (Steinmüller, 1997) (see Figure 8.2). But ultimately, the future is not predictable, i.e. it cannot be planned in any determinist sense (Cuhls, 2000). Similarly scenarios are at best mathematical or logical constructs of present assumptions, and hence need to be looked upon very critically, because – as experience shows – such scenarios, even when seemingly perfectly calculated, have a very high error rate.

Forward-scenarios, or also the ordinary trend projections of the early 1990s that sought to project the role of renewables in the electricity mix, came to much lower percentages for today than are already reached. But for the planning of 100 per cent renewable energy regions scenarios are very important, in the form of traceable, precisely yet imaginatively formulated and visualized futures and development paths leading to these. While acknowledging their limitations, the method of calculating scenarios can be used as a planning technique from which strategic decisions derive.

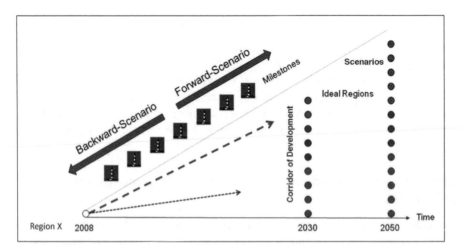

Figure 8.2 Path of development for potential 100% renewable energy regions

Source: Moser (2008a. 2008b)

As a central and vivid future vision for a general description of a '100 per cent renewable energy region', the research team designed a positive, desirable mission statement of an 'ideal region' that can be defined as follows:

> A 100% renewable energy 'ideal region' covers its energy demand entirely with renewable energies, is based on very high levels of energy efficiency and includes the regional renewable potentials comprehensively. The energy supply is ecological, sustainable, secure and increases the regional added value. Regional players are involved and a high acceptance for this way of energy supply exists in the regional population. Regional key players have framed together with end users, producers and practitioners the development process, which leads to a comprehensive energy supply with renewable energies. For cost reduction and safeguarding the security of supply they cooperate in a network with other ideal regions. Energy efficiency, sustainable energy production and regional activities for an energy-conscious behaviour are integral elements in the region. (DeENet, 2009)

A general mission statement such as this needs to be adapted to respective regions with their specific local conditions.

Dimensions of an integrated overall model of 100% renewable energies

Establishing 100 per cent renewable energy regions requires a holistic approach, because regional energy policies concern a variety of institutions, especially when the three sectors of electricity, heat and transportation are considered.

Offering a first overview of different approaches, levels and cross-references between the given dimensions, Moser and Hoppenbrock (2008) developed a system model for renewable energy regions (see Figure 8.3) that comprises different topics:

- thematic dimension (economy, techniques, policy, sociology, ecology, markets etc.);
- spatial dimension (building, quarter, village, town, district, region etc.) (see Figure 8.1);
- normative dimension (100 per cent target, sustainability, regional development etc.);
- temporal dimension (past, present, future, long-term future).

A decentralized production of renewable energies leads to many small units that have a big impact on the landscape, and thus effects the perception of the inhabitants. Decision-makers not only focus on one goal, but usually refer to a variety of objectives such as regional development, jobs, CO_2 decline etc. All in all, the transformation of the energy system might turn out to be a task for a whole generation. By considering the temporal dimension, the process dynamic, the genesis, the latest developments and future visions are included.

It is essential to disclose that partial models and an isolated approach to the subject are not sufficient to enlighten the interrelations due to the complexity of the structures. Hence integrated systemic considerations are necessary that combine the individual sub-models with an overall model. This approach also emphasizes the necessity of back-casting scenarios and other planning methods (see Figure 8.3).

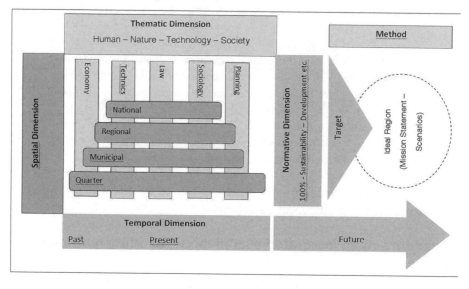

Figure 8.3 Integrated model for achieving an 'ideal region'

Source: Moser and Hoppenbrock (2008)

Guideline: How to achieve 100% renewable energy regions

Based on this model and further practical field experiences, it is possible to outline criteria and critical factors that seem to be necessary or beneficial on the way towards a 100 per cent renewable energy region. The suggested guideline outlines regional fields of activity without ignoring the interdependencies pointed out above. The following propositions are each backed up with best-practice regions found among a sample of regions (see Figure 8.4). Each of these regions is unique due to its natural resources, but also in its mindset, its history, the political landscape etc. Therefore there is no standard to meet or a special way to go. The guidelines include:

- Board decisions aiming at 100 per cent – usually municipalities in Germany are enacting so-called 100 per cent resolutions on different administrative levels. These are not only letters of intent, but comprise measurable objectives, timetables and also financial resources.
- Exploration of renewable resources – evaluate the sustainable and accepted renewable energy potentials, the efficiency potentials and possibilities of

energy saving, and calculate the (previous and future) generation capacity and demands. Estimate the chances for realizing a regional 100 per cent target by using different technologies. Setting the spatial range can already be crucial to the process – a 'minimum size' is necessary for the layout of 100 per cent-targets.

- Spatial planning – after evaluating the regional potential for renewables it will be necessary to integrate the increase of renewable energy aspects into spatial planning.
- Pioneers and personalities – identifiable and competent persons take responsibility and push the process. Along with local decision-makers, 'the early adopter' or committed amateurs can play a decisive role as opinion leaders promoting renewable energies.
- Added value – decentralized energy technologies, especially those financed with regional capital and by different stakeholders, accumulate the added value of a region. Especially among politicians and people outside the ecological movement, this argument is a strong motivation to promote renewable energies.
- Participation – regional inhabitants should be highly integrated into a preferably transparent development process to achieve a high degree of acceptance and team spirit and to avoid resistance to renewable energy implementations. Besides good publicity, financial participation (civil companionships) should also be taken into consideration.
- Entrepreneurship – the transition of the energy system depends on entrepreneurship in the energy markets; regional or local administrations can usually not be counted as economic players themselves. Thus it is important to foster start-ups in the energy sector as well as their counterparts – small companies that often emerged from agriculture or crafts etc. Some German regions have successfully incorporated public–private partnerships in order to be able to fulfil the role as municipal energy suppliers, or to organize projects and plants.
- Beacon projects – the realization of single plants makes renewable energies more tangible and establishes models that sometimes serve as a kick-off for other projects, even in other regions. In addition they reduce prejudices and sensitize the population to the implementation of renewable energies into the landscape.
- Master plan – development and implementation of a master plan (roadmap) with measures, interim goals and strategies. The unpredictability of the future at this point allows a departure from strict scientific methods. In order to develop concrete recommendations and decisions for the conversion of the energy system based on renewable energies, creativity supporting, intuitive planning methods of the futurology are reasonable (see Figure 8.2). The adjustment of strategic planning is also necessary in many existing areas, such as rural planning, agriculture or academia.

Examples of 100% renewable energy regions

Fully supplying regions with renewable energies has not yet been achieved in Germany, not in any particular region, nor in any particular market (electricity, heat or transportation). Still, a number of promising regional approaches have emerged. Figure 8.4 shows an already impressive number of regional processes that are currently of special interest to the research team. The regions on the map are comparable only in the domain of defining objectives (100 per cent more or less) and/or agenda setting. For instance, 100 per cent electricity at the coastline of the North Sea seems feasible by wind energy, but solutions for the heat market are still missing. The southern districts around Munich have neither the potential nor the public acceptance for promoting wind energy, but are focused on solar energy or bioenergy. Currently, 5.6 million people that live in these regions are represented by a 100 per cent target and their number is still growing.

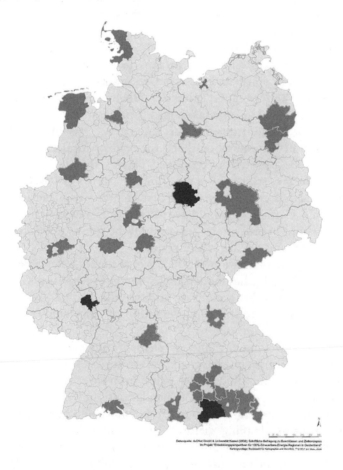

Figure 8.4 100 per cent renewable energy regions in Germany (December 2008)

Source: deENet (Hrsg.) (2009)

The following three regions (marked in black in Figure 8.4) are very different in their approaches to achieving the changeover to renewables, but each of them has got a 100 per cent resolution and serves as a good example of the ongoing process in Germany.

Model region Harz, Druiberg (2008)

The small town of Dardesheim in the Harz is in terms of figures already producing 10 times more energy for electricity, heating and transport than the 1000 inhabitants of the town consume – some 40 times its electricity demand alone (Uken, 2007). One third of household electricity is provided by ten large photovoltaic rooftop systems. In addition tiny Dardesheim features the world's largest wind turbine: a 6MW, 125m giant. The municipality also operates some electrically powered cars, fuelled with solar and wind energy, and the local garages offer vehicle modifications from diesel fuel to vegetable oil.

With these implementations, Dardesheim has served as a nationwide beacon and is now the centre of the first attempt to realize the 'Combi-Plant' in a defined region, named 'model region Harz'. Together with nearby windparks, biogas plants, a pumped storage hydropower plant and another windpark with a capacity of 20MW (already approved), the supply of most of the Harz region will be ensured by purely renewable energy production. Further aims are to increase the regional productive capacity, create additional jobs and stimulate investment by additional EEG-based regional suppliers, to strengthen the employment base and promote tourism.

Energy turnaround in the Oberland region, Oberland (2009)

The region of Oberland is situated in the deep south of Germany, between Munich and Innsbruck in the centre of the Bavarian Alps. In order to achieve a full supply with renewable energies within the next 30 years, this region established a civic foundation and is therefore a good example of a participation model with strong regional players. This foundation with the name 'Energy Turnaround Oberland' was recognized on the 7 November 2005 by the federal state government of Upper Bavaria as a non-profit organization supported by 129 sponsors. The counties Bad Tölz-Wolfratshausen and Miesbach support it with a total of 20,000 members, and resolved to generate its entire electricity demand from renewables by 2035. Almost 40 municipalities in these two counties participate in the foundation projects. All municipalities, which have signed and will sign the resolution, commit themselves to increasing renewable power and energy efficiency, in a drive to enhance municipal independence and community engagement. The slogan 'Energy Turnaround Oberland' aims at full energy independence from energy imports no later than 2038. The charter of the foundation also prohibits the long-distance import of alternative energy such as bioethanol, palm oil or maize-based fuel. The aimed-at turnaround is to be comprehensive, and focuses on electricity as well as the heat and the transport sectors. At present the following projects are being pursued:

- public solar power plants on public roofs, such as school buildings;
- a campaign for the insulation of single-family houses for energy saving (in the private and public sector) and to increase wood chip, pellet and biogas plants with CHP plants;
- guidance for municipalities to utilize the waste heat of deep geothermal energy plants.

County Alzey-Worms: 100% renewable energy in 13 years

The county of Alzey-Worms covers an area of nearly 600km², is home to around 125,000 people and consumes approximately 800 million kWh of electricity each year. With support from energy service company juwi Holding AG (that also researched renewable energy resources), it aims to produce its entire electricity demand by 2020 from renewable sources. This target is pursued in cooperation with a local association of municipalities and communities as well as with the government, businesses, organizations and inhabitants of the county. With the financial and organizational assistance of juwi Holding AG, the construction of five two megawatt wind turbines, three 2.5MW wind turbines and two 500kW bioenergy plants is planned. At the same time, a large-scale photovoltaic installation of 5MW is being prepared and the expansion of private rooftop systems, also with a capacity of 5MW, will be pushed. Significant efficiency increases and savings in the electricity consumption of 12.5 per cent by 2020 are part of this plan (www.juwi.de).

References

Cuhls, K. (2000) *Wie kann ein Foresight-Prozess in Deutschland organisiert werden?*, Friedrich-Ebert-Stiftung, Bonn

deENet (Hrsg.) (2009) Schriftliche Befragung von Erneuerbare-Energie-Regionen in Deutschland - Regionale Ziele, Aktivitäten und Einschätzungen in Bezug auf 100% Erneuerbare Energie in Regionen

DeENet (ed) (2009) *Development Perspectives for Sustainable Renewable Energy Regions in Germany*, www.100-ee.de

Kröcher, U. (2007) *Die Renaissance des Regionalen. Zur Kritik der Regionalisierungseuphorie in Ökonomie und Gesellschaft*, Westfälisches Dampfboot, Munster

Moser, P. (2008a) 'Bioenergiedörfer – Es fehlt an Effizienz', *Sonne, Wind & Wärme*, vol 18, pp78–79

Moser, P. (2008b) 'Entwicklungsperspektiven für nachhaltige 100%-Erneuerbaren-Energie-Regionen in Deutschland', Lecture at the Berlin Energy Days, 5 May, www.100-ee.de

Moser, P. and Hoppenbrock, C. (2008) 'Modelle und gesellschaftliche Prozesse für ein regionales Energiesystem', in Bonow, M. and Wolfgang, G. (eds) *Regionales Zukunftsmanagement* 2, Springer, Berlin/Heidelberg, pp72–85

Moser, P. and Meyer, B. (2002) 'Szenarienentwicklung und -operationalisierung für die suburbane Kulturlandschaft', UFZ-Report Nr. 21/2002, Helmholtz-Zentrum für Umweltforschung, Leipzig

Rahmstorf, S. and Schellnhuber, H. J. (2007) *Der Klimawandel. Diagnose, Prognose, Therapie*, 5th edition, C. H. Beck, Munich

Scheffer, K. (2008) 'Vom Bioenergiedorf zur autonomen Solar-Region', *Solarzeitalter*, vol 4, pp23–30

Steinmüller, K. (1997) *Grundlagen und Methoden der Zukunftsforschung*, SFZ-WerkstattBericht Nr. 21, Sekretariat, für Zukunftsforschung Gelsenkirchen

Tischer, M., Stöhr, M., Lurz, M. and Karg, L. (2006) *Auf dem Weg zur 100%Region. Handbuch für eine nachhaltige Energieversorgung von Regionen*, B.A.U.M. Consult, Munich

Uken, M. (2007) 'Die Harzer Stromrebellen', *Die Zeit* 30 October, www.diezeit.de

Chapter Nine

Renewable Regions:
Life After Fossil Fuel in Spain

Josep Puig i Boix

On the day of the 2007 Summer Solstice, a group of Catalan NGOs made public a study on 100 per cent renewable electricity supply for Catalonia (Doleschek et al, 2007), the north-eastern region of Spain. The authors of the study showed that to switch the present electricity supply system in Catalonia to a renewable one would be within reach if appropriate policies were implemented. Also, some time before, Greenpeace Spain had undertaken a similar study for Spain as a whole (Greenpeace Spain, 2007).

But despite these inspiring studies, the Spanish energy situation is not going in a good direction: increasing energy demand, lack of efficiency, CO_2 emissions out of control, mounting energy dependency on fossil fuels, to name just some of the major problems. To make matters worse, the Spanish government never took the responsibility of asking for a study to assess the challenges of supplying the country's electricity from a 100 per cent renewable system.

Why speak about a 100 per cent renewable energy supply? The answer is simple: the present energy situation is unsustainable because it is based on non-renewable sources that generate energy waste and emissions on a massive scale, both in CO_2 and nuclear radiation. Time is running out for shifting the current energy path to a renewable one. Spain's quandary epitomizes the paradox of many countries that utterly depend on oil and uranium energy imports and fail to comply with the Kyoto Protocol, while sitting on abundant renewable resources: sun, wind, biomass, hydro, or in the words of Samuel Taylor Coleridge (1898), 'Water, water, everywhere, Nor any drop to drink'.

The current situation of renewable energy in Spain

Before the end of the 1990s, Spain began to deregulate energy markets and adopted some national policies to develop renewable energies. A number of big steps have been taken since then. The Law of the Electricity Sector (Ley 54/97) fixed that at least 12 per cent of all primary energy should be covered by renewables. The Renewable Energy Promotion Plan 2000–2010 was adopted at the end of 1999 and fixed a renewables target of the equivalent of 9.5 mtoe for the year 2010. The updated Renewable Energy Plan 2005–2010 (Plan de Energia Renovable (PER) 2005–2010) was adopted in August 2005 and increased the target for 2010 to 10.5 mtoe.

According to the Spanish Energy Agency (IDAE), at the end of 2007 renewable energy technologies provided 7.1 per cent of primary energy consumption (3.1 per cent biomass, 1.6 per cent hydro, 1.6 per cent wind and the remaining 0.8 per cent distributed between biofuels, solar thermal and solar PV, geothermal and municipal solid waste. Regarding electrical energy, the contribution for renewables was 20.5 per cent (9.8 per cent hydro, 9.0 per cent wind and the remaining 1.7 per cent distributed between biomass, biogas, solar PV and municipal solid waste).

If we take a look at each renewable energy source the situation at the end of 2007 was as follows:

* Wind energy ranks first with more than 15,000MW installed at the end of 2007, with a 30.3 per cent increase from 2006. The goal for 2010 (20,155MW) will be easily reached and perhaps exceeded.
* Solar thermal, with almost 1 million square metres (700MWth) installed, the goal set by PER of 3.5 million square metres in 2010 will not be reached, despite the fact that more than 50 municipalities around Spain have adopted Solar Energy Ordinances (City of Barcelona, 1999). This is also despite the inclusion of a solar thermal obligation in the new Spanish Building Technical Code (CTE – Código Técnico de la Edificación).
* Solar photovoltaics, in contrast to solar thermal, has experienced rapid development in the last two years, reaching 554MWp at the end of 2007 (in contrast to 27MWp installed at the end of 2003). In 2007, solar photovoltaics exceeded the goal of 400MW fixed by PER for 2010.
* Solar thermoelectric or CSP is also booming due to very attractive premiums for developers. The first commercial power plant was operational at the end of 2007 (PS-10, 11MW, near Seville), with the second one (PS-20, 20MW) producing electricity at the end of 2008. Both are based on tower technology, but the group building them, Abener, has already started construction of two projects based on parabolic troughs of 50MW each (Solnova 1 and Solnova 3). Dozens of other projects are being planned and building started; among them, Andasol 1 (feeding electricity to the grid at the end of 2008) and Andasol 2 are the more advanced.

All these developments have been implemented in line with FiT policies since the 1990s (RD 2828/98), and based on some policies adopted in the early 1980s (Ley 82/1980). But why have some Spanish regions developed renewable energy assets while others have not, despite the fact that national policies apply for all regions (RD 436/2004)? For example, and referring to wind (see Table 9.1), some regions such as Castilla-La Mancha had more than 3000MW at the end of 2007 (or Galicia and Castilla y León with almost

Table 9.1 Wind power in Spain

Regions	Surface	Population	Density	1998	1999	2000	2001	2002	2003	2004	2005	2006	2007	kW/km²	W/hab
	km²	2006	hab/km²	MW	MW	MW	MW	MW	MW	MW	MW	MW	MW		
Andalucía	87,598	7,975,672	91	115	127.8	150.3	158.1	163.63	233	361.63	448.24	606.56	1,459.71	17	183
Aragón	47,720	1,277,471	27	128	208.5	230.4	404.3	733.93	995	1,206.94	1,407.14	1,532.44	1,723.54	36	1,349
Asturias	10,604	1,076,896	102	0	0	0	24.4	73.72	121	146.00	164.01	198.86	277.96	26	258
Baleares	4,992	1,001,062	201	0	0.2	0.2	0.2	0.20	0	3.65	3.65	3.65	3.65	1	4
Canarias	7,447	1,995,833	268	80	81.8	114.7	120.4	126.92	128	129.49	129.49	133.24	133.24	18	67
Cantabria	5,321	568,091	107	0	0	0	0	0.00	0	0.00	0.00	0.00	17.85	3	31
Castilla y León	94,225	2,523,020	27	16	122.2	228.2	352.9	634.93	924	1,523.17	1,816.87	2,122.91	2,818.67	30	1,117
Castilla-La Mancha	79,462	1,932,261	24	0	111.9	348.2	493.2	741.17	1,010	1,585.50	2,017.66	2,281.46	3,131.36	39	1,621
Catalunya	32,113	7,134,697	222	20	58.8	70.7	83.4	86.36	86	94.37	143.87	225.3	347.44	11	49
Comunidad Valenciana	23,255	4,806,908	207	0	2.8	2.8	2.8	20.49	20	20.49	20.49	333.99	590.94	25	123
Extremadura	41,635	1,086,373	26	0	0	0	0	0.00	0	0.00	0.00	0.00	0.00	0	0
Galicia	29,574	2,767,524	94	232	438.1	600.5	937.7	1,314.99	1,579	2,102.21	2,369.28	2,619.64	2,951.69	100	1,067
La Rioja	5,045	306,377	61	0	0	24.4	73.9	203.52	272	346.87	408.62	436.62	446.62	89	1,458
Madrid	8,028	6,008,183	748	0	0	0	0	0.00	0	0.00	0.00	0.00	0.00	0	0
Murcia	11,313	1,370,306	121	6	6	11.22	11.22	11.22	32	48.97	54.97	67.72	152.31	13	111
Navarra	10,390	601,874	58	237	318.4	467.8	554.1	690.51	717	849.86	899.36	916.36	937.36	90	1,557
País Vasco	7,235	2,133,684	295	0	0	24.5	27	26.90	85	84.77	144.27	144.27	152.77	21	72
Ceuta	19	75,861	3,993	0	0	0	0	0	0	0.00	0.00	0.00	0.00	0	0
Melilla	13	66,871	5,144	0	0	0	0	0	0	0.00	0.00	0.00	0.00	0	0
Total	505,989	44,708,964	88	834.00	1,476.50	2,273.92	3,243.62	4,828.48	6,203	8,503.92	10,027.92	11,623.02	15,145.11	30	339

Source: Based on data from Asociación Empresarial Eólica

3000MW), while Catalonia had only 350MW? This is despite the fact that the first Spanish grid-connected wind generator was designed, constructed and erected in Catalonia in the early 1980s. Why has Catalonia not been able until now to develop its wind potential, despite Spanish national renewable energy policies that apply to the region?

In order to answer this question, it is instructive to learn from past developments that have led to the present situation in Catalonia and compare these with the case of Navarra.

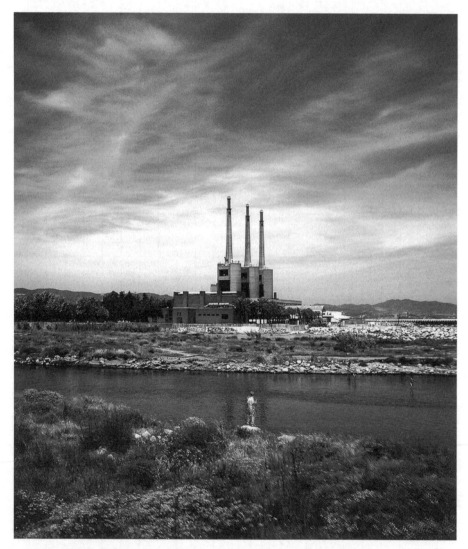

Figure 9.1 Barcelona's San Adrian power plant: The plant displays a sleek design, but despite its much-touted credentials is still a relatively large polluter, emitting 513,193 tons of CO_2 annually at an output of 1.2TWh

Source: Juan Antonio Zamariipa Esqueda

The case of Catalonia

Catalonia is the north-eastern region of Spain, with more than 7 million inhabitants (15.5 per cent of the Spanish population) living in 32,113 km² (6.4 per cent of the Spanish area) with 222 inhabitants/km². In 2007, more than 50 per cent of all the electricity it consumed was generated by three nuclear power reactors (in the mid-1980s nuclear power provided more than 80 per cent of all the electricity to Catalonia).

The first proposal for a renewable-based energy supply for Catalonia was published in the early 1980s (Corominas and Puig, 1982) by people involved with Ecotècnia. This study was inspired by a number of pioneering studies (Lovins, 1977; Groupe de Bellevue, 1978; Todd, 1979; Morris, 1979; Taylor, 1979; Leach et al,1979; and Lönnroth et al, 1980) carried out in the late 1970s by different groups in various countries.

In Catalonia, two examples show the impact of clear commitments by local or regional governments to achieving results. The first concerns wind technology and the second solar thermal energy. The case of wind involves a private company, Ecotècnia cooperative, that has been successful in developing wind technology since the early 1980s. The case of solar thermal energy focuses on the city of Barcelona that has been leading the process to adopt the first world municipal solar thermal ordinance, having made it mandatory since 2000 that all new and integrally refurbished buildings have to have solar water heating systems. The full story of Barcelona Solar Ordinance has been already published (Puig, 2008), but now details of the Ecotècnia case are published here for the first time.

2008 will be remembered in the wind history of Catalonia as the year in which the Catalan cooperative (pioneer of wind turbines design and manufacture) Ecotècnia, integrated the international group Alstom. Almost three decades had passed since a group of nine persons with high technical qualifications, committed to environmental thought and the practice of alternative technology, started to work to make possible the birth of a cooperative with the goal of developing technology for using renewable sources of energy.

It was in the late 1970s, within the framework of the energy crises and in the constructive fever of nuclear power stations and multinational companies looking for uranium in Catalonia, when a series of circumstances lead the first people interested in the renaissance of the wind energy in Catalonia to get together. In 1978, a graduate wind project (Meseguer, 1978) was presented to the Engineers School of Barcelona. During the month of June, Navarra's Association of Architects organized a special event in Pamplona about alternative energy (Autores varios, 1979), where Tvind School representatives (from Jutland in Denmark) attended and explained their story of designing and constructing a megawatt-size wind turbine with three blades (each 27m long) on top of a 53m tower. Several members of the group, Tecnologías Alternativas Radicales y Autogestionadas (TARA) (TARA, 1977), participated in this conference and at the end of 1978, a Barcelona underground magazine (Alemany, 1978) published an article explaining how the Danish Tvind schools constructed and put into operation a megawatt-size wind turbine.

On 21 November, the Technical Commission of Energy of the Association of Industrial Engineers of Catalonia was set up. Dr J. Corominas and I attended its founding ceremony as both of us had been very active in initial work of a wind energy subcommittee. From February 1979 until July 1984, I acted as coordinator of this subcommittee and its members went on to found Ecotècnia.

The wind energy subcommittee was very productive, for example, it produced a paper (Puig, 1979) that was presented at the Catalan Conference of Engineering, proposing a wind energy research plan, including a study of Catalonia's wind potential and even the testing of existing wind machines and the development of wind technology. The subcommittee organized the first Conference on Wind Energy in January 1980 (Comissió Tècnica d'Energia, 1980), where the engineer Lucien Romani was invited to explain the French experience in wind technology, especially the experience of the 800kW Best-Romani wind turbine that worked in Nogent-le-Roi (France) from 1958 to 1963, funded by EDF. Also a local manufacturer (Aerogeneradores GEMZ) was invited, as was the builder of a pumping wind machine. The subcommittee organized the first wind energy course in June 1980 (CPE-AEIC, 1980) with members of the subcommittee as teachers. It produced several papers· (de Cisneros, 1982; Departament d'Indústria i Energia de la Generalitat, 1981) on wind energy that were presented at the Conference of Industrial and Energy Policies and to the Conference on Studies and Technical Proposals for the development of the technology and energy policies of the government of the Generalitat de Catalunya (January 1981).

In this context a group of people started in the autumn of 1980 to nurture the idea of creating a cooperative for the design and manufacture of wind turbines. This became Ecotècnia. Ecotècnia was the culmination and realization of a long process of debate about the relationship between energy, technology and society. The group began to form in the late 1970s, involving people who had been organizing courses, giving lectures and writing features in magazines about energy alternatives to the official energy policies based on the country's push to nuclear power. In an article published in a Spanish magazine, it was announced publicly that the Cooperative for Technological and Energy Autonomy (CATE) would be set up (Alemany and Puig, 1980).

The debate about the creation of a cooperative structure to develop technology for the use of renewable energy sources started in the autumn of 1980 and was inspired by authors such as Schumacher (1975), Dickson (1975), Bosquet and Gorz (1977), Lovins (1979), the publications of the Open University's Network for Alternative Technology and Technology Assessment (NATTA) and the magazine *Undercurrents* (Harper and Boyle, 1976).

In May 1980, the group of pioneers of Ecotècnia started making contact with the Spanish government body in charge of technology development and innovation, the CDTI (Centro para el Desarrollo Tecnológico e Industrial) to explore the possibilities of obtaining funding to carry out the technological development of a wind turbine. In January 1981, they presented a first proposal but this was not accepted.

The formal constitution of Ecotècnia cooperative was held in the city of Barcelona, on 2 April 1981. The founding capital was 80,000 Spanish pesetas (about €500), coming from 8 people. Ecotècnia's mission was to offer a series of products and services to empower, develop and promote a technology within the reach of anybody, that would provide more autonomy to workers and users, allow a better use of local resources and would be more respectful of the environment, encouraging the use of renewable and non-polluting energies that would not artificially bloat prices due to market restrictions, or because of inferior management, and that would be open to participation and that would be inclusive and egalitarian in the organization of the work needed.

The proposal submitted to the CDTI was ultimately not accepted but nonetheless it had consequences: in August 1981 CDTI announced a competition for the design of a wind turbine prototype of 5–10kW of power, framed under the Spanish Innovation Plan that had been funded by the Spanish government and the CEOE (Confederación Española de Organizaciones Empresariales) (Ministerio de Industria y Energía y CEOE, 1981). Ecotècnia started to work on its prototype in September and was one of the four winning groups, together with Gedeón S.Coop., STS S.Coop. and IDE.

While waiting for the result of the competition, I joined in the 1981 European Wind Energy Study Tour of the US, the first expedition of European wind experts to developments in the US, especially such pioneering outposts as NASA's Lewis Research Center, the Wind Test Plant in Rocky Flats, the nascent wind farms in California and in the state of Washington.

The winning of the CDTI competition led to much pomp and ceremony, but despite the signing of an agreement with the Energy Ministry's General Direction of Industrial and Technological Innovation, the contractual document never made it back into the hands of the winners of the competition, and neither did the prize money (€84,000). In order to avoid a public outcry, the CDTI negotiated the funding of 90 per cent of the development cost of the wind turbine prototypes with the four winners. Finally the collaboration agreement with the CDTI was signed in July 1982, which made it possible to implement the projects.

Two significant events accompanied this process: the publication in April 1982 of the first Spanish book on wind energy technology (Puig et al, 1982), starting a series on alternative technologies directed by Ecotècnia, and second, the first doctoral thesis about wind energy (Puig, 1982).

Finally the Ecotècnia 12/15 wind turbine was installed and connected to the grid in the small rural town of Valldevià (Municipality of Vilopriu, Comarca de l'Alt Empordà, Girona province, Catalunya). The wind turbine was connected somewhat outside the legal context, since then there were no regulations in place. It was a three-bladed wind turbine, each blade 6m long

(12m diameter), mounted on a 10m steel frame tower with 15kW of rated power.

On 10 March 1984, the official inauguration happened, with a big party on site. The party started at Vilopriu Town Council with a presentation of the cooperative and of the project, accompanied by an ambitious slideshow on wind energy and its role in modern society. Afterwards, the 500 guests moved to the site of the turbine to officially switch it on, not without the obligatory fireworks. TV coverage allowed everybody to witness the birth of modern wind technology in Catalonia, and both national and local press reported the inauguration of the renewable age.

For three years, Ecotècnia's 12/15 wind turbine was a true test case for all the components. The prototype was submitted to the Second Creativity's Award of the industrial engineers, and was rewarded with an honourable mention (with the jury still preferring the wonders of the fossil fuel era for first prize: a diesel locomotive designed for train shunting).

During the delivery ceremony of the prize in June 1984 in a Barcelona downtown hotel, when receiving the award from Catalonia's president, Ecotècnia quoted a previous presidential speech: '*No podemos volver a molinos de viento*' ('We can not return to the times of windmills') (*Noticiero Universal*, 29 May 1982) and spoke of the Californian example of building wind farms.

Ecotècnia 12/15 wind turbine was reported in some international wind technology magazines. The newsletter of the European and UK Wind Energy Associations (*Windirections*, 1984) put a picture of the Ecotècnia 12/15 wind turbine on its front page and reflected it as a major event in wind energy technology in Europe. Also, *World Wind* (1984) magazine published a full-page article where it covered the inauguration and explained the story of the competition. Also, in one of the first wind energy conferences, EWEC '84 (European Wind Energy Conference and Exhibition, Hamburg), organized by the EWEA (European Wind Energy Association), a paper was presented on the evaluation of Catalonia's wind potential and the development of a 15kW wind turbine (Puig and Corominas, 1984).

The Ecotècnia 12/15 wind turbine was the basis from which the first generation of Catalonia's wind machines started to be manufactured (12/30 series), of which Ecotècnia sold and installed 29: 2 in the Castilla-La Mancha region, 4 on a Spanish commercial wind farm, 20 in the Ontalafia (Albacete) and Tarifa (Cadiz) wind farms and 3 more in Los Llanos, Figuerola del Camp and Roses. All these wind farms were included in the first Spanish Renewable Energy Plan (Plan Energético Nacional 1991–2000).

The accumulated experience, made it possible for Ecotècnia to undertake the development of more powerful wind turbines, rated at 150kW. The first prototype, designed and built with the support of the IDAE, was installed at Tarifa, the most productive wind machine (kWh/m2) in Europe. With this machine the second generation of wind energy systems was initiated, which would be followed by the generations of 225kW, 640kW, 750kW, 1250kW, 1670kW and 3000kW (the official inauguration of the first 3 MW wind turbine took place on 25 July 2008 in Les Colladetes, with the presence of Catalonia's President).

Catalonia started pushing wind technology with almost no support from the Catalan government. It took more than 15 years to see the results of the hard work done by people engaged with the idea that a modern society could be supplied by renewables.

Now, more than 25 years later, the Catalan government continues to avoid making any clear commitments regarding renewable energy sources. For example, its recent Decree for the regulation of wind and solar photovoltaic developments (Generalitat de Catalunya, 2008) was strongly opposed by the main environmental organizations (Greenpeace, Ecologistes en Acció and Eurosolar Spain).

The case of Navarra

The small Spanish region of Navarra (1.3 per cent of Spanish population and 2 per cent of the Spanish land area) with 58 inhabitants/km², became over a short period of time a leader in renewables (mainly wind energy) in Spain.

The process started in 1989 when the Navarra government decided to start working to make the region a leader in renewable energy. The first steps were: to assess the region's wind energy potential and to create a small company, EHN to develop small hydro sites in the region (more than 100 sites with a power capacity of 195MW). EHN built its first wind farm (El Perdón) 15km from the capital city of Pamplona in 1994. In the same year, the Navarra government, through SODENA and with EHN and Vestas, joined forces to set up Gamesa Eólica to supply wind turbines to EHN.

In 1995, Navarra adopted its 1995–2000 Energy Plan with a goal to reach 341MW of renewable energy capacity by 2000. The objectives of the Plan were easily fulfilled and the year 2000 ended with 667MW of renewable capacity installed, of which 474MW were from wind energy. At the end of 2006, the total renewable power capacity installed in Navarra was 1164MW, of which 941MW were from wind on 32 wind farms with 1164 wind turbines.

Navarra is the location of several wind farm developers (Acciona Energía, Eólica Navarra, Gamesa Energía, etc.) and wind turbine manufacturers (Acciona Windpower, Ecotècnia, Gamesa Eólica, Ingeteam, etc.). Also this region has attracted leading national centres specialized in renewables (CENER and CENIFER).

Navarra is also leading solar photovoltaic development, since EHN teamed in 1997 with other partners to create AESOL. This company developed the innovative concept of '*huerta solar*' or solar farm involving ordinary citizens in solar investments. By September 2008 it had developed 18 *huertas solares*, with a power capacity of 61.5MW, involving more than 3500 people, mainly in Navarra but also in Castilla-La Mancha, Extremadura and Aragon regions.

The strong regional political commitment has made the region of Navarra a shining example in renewable energy development. The policy context made possible the formation of a close partnership between regional government, private companies and financial institutions, and led to results in a short period of time.

The results

From the above the differences between Catalonia and Navarra are evident, despite both regions being subject to the same Spanish legislative framework.

The main difference is the commitment of regional governments: in Navarra a strong commitment to develop renewable energy policies, in Catalonia no real commitment at all.

The main conclusion of these stories is that local and regional governments have to play an active role in developing renewable energies. Without active renewable energy policies at regional or local levels and despite the active involvement of other local actors and active policies at a national level, we will never see a society based on 100 per cent renewable energy.

Another difference between these two regions is the question of energy democracy. In Navarra a private company has created and developed a new concept, enabling the ownership of renewable energy systems by ordinary people. In Catalonia, no private company has followed that path. Only a local NGO, Fundació Terra, started in 2007 to pursue the idea in Barcelona, without any support from the city government.

The Solar Catalonia proposal

To reiterate, a pathway to a 100 per cent renewable electricity system for Catalonia was proposed in July 2007 by a group of Catalan NGOs. The main goal of the proposal was to pressure the regional government to change its regional energy policies.

The objective of the Solar Catalonia study is to show that the region would be able to supply its own need for electricity from renewable sources. This fact-based vision of a future energy supply is very important to influence the discussion about the change from fossil and nuclear energy sources towards a sustainable energy system, especially, as the ongoing discussion regarding the possibilities of renewable energy and efficient design has been negatively influenced by data on the availability and potential of these technologies.

The goal of the project is to show that a sustainable, renewable and efficient energy system is capable of supplying Catalonia's current needs. The study does not assume any major changes in lifestyle, living standards or demographic composition. There are no assumptions regarding future economic development in terms of GDP or the like.

Although Catalonia has shown strong economic growth over recent years, it did not perform well with regard to energy intensity. It is quite clear that energy intensity in the Catalonian economy must be reduced in order to shift to a sustainable energy supply and to make a contribution to climate protection. The scenarios within the work highlight a development towards halving electricity intensity in the three most important sectors of electricity consumption until 2050. This means making great efforts to improve the efficiency of electricity use, but the authors are convinced that this is feasible from a technological point of view. Further technological development towards more efficient appliances will assist such a development and in the restructuring of our economies. Redefining the relationship between energy consumption and wealth may be necessary but, in the end, climate change and its serious consequences will force us along this path. One fact is quite clear: we have to start now in order to keep the transition smooth and to avoid the most serious consequences of climate change.

Taking this proposed course of action will lower Catalonia's electricity consumption to the 1993 level until 2025 and to half of 2003's electricity consumption by 2050. Although further reductions will be harder to achieve the further we step into the future, a certain level of energy intensity will remain. Reducing energy intensity by half sounds very difficult, yet this means only undoing the increase experienced between 1993 and 2003. The remaining effort of efficiency improvements does not seem like an insurmountable goal.

Two scenarios show the feasibility to achieve a fully renewable supply, one until 2035 (Fast Exit Scenario) and the other until 2045 (Climate Protection Scenario). This is not a matter of potential but of setting and pursuing ambitious goals, encouraging policy and people and – of course – the financial investments Catalonia and it's people are willing to take. The scenarios show that the financial aspect is not that big an obstacle as one might expect. With an annual investment into renewable capacities peaking at €104 (2006 value) per inhabitant in the Fast Exit Scenario and €85/capita in the Climate Protection Scenario, the financial burden to achieve a clean, climate-friendly electricity supply in Catalonia is moderate in the authors point of view; in 2030 investments would be €103/capita in the Fast Exit Scenario and €68/capita in the Climate Protection Scenario.

These financial figures reflect the peak investments during the whole development considered in the study. The calculation of the average annual payments for the two different scenarios results in €58 per inhabitant a year in the Climate Protection Scenario and €84 per inhabitant per year in the Fast Exit Scenario.

Compared to the Catalonian GDP (€181,029 million in 2005) the annual costs of the scenarios are 0.2 per cent of GDP for the Climate Protection Scenario and 0.3 per cent for the Fast Exit Scenario, on average.

Any energy supply system must guarantee sufficient production and distribution of electricity, heat and fuels to meet the demand for energy at any time throughout the year, usually using different energy conversion technologies. Energy is supplied in the form of electricity, heat or fuels, with heat and fuels having the advantage that both can be stored for later use and can be easily transported. So it is not necessary to consume heat and fuels immediately or directly at the production site. Heat can be stored in thermal reservoirs and distributed via DH networks. In contrast to heat and fuels, which dissipate with time – thus setting a limit on storage time and distribution distance – fuels from biomass or hydrogen do not have quite this limitation in storage time or in transport (depending on the fuel type – solid, liquid or gaseous), although some storage losses must be considered here as well.

The situation is completely different with electricity. The necessity of producing enough electricity, on demand and on time, makes this type of energy the most critical component in an energy supply system. While electrical transport via the public grid is quite unproblematic, storing electricity directly on a large scale is material and cost intensive. Also, storage in batteries and accumulators can involve the use of toxic substances. Therefore this option is not considered in the study as it is not appropriate for a sustainable energy supply system. Indirect storage can be used, for example, pumped hydro storage systems.

An energy supply system that is based almost completely on renewable sources increases the focus on timely energy dispatch and supply due to the fluctuating nature of some renewable energy sources, such as solar and wind. Including such fluctuating sources into the public electricity supply means that the power produced by those sources might decrease relatively fast. Of course electricity production from fluctuating sources can be estimated by weather forecasting but a portion of uncertainty still remains. Fortunately, there are other renewable technologies with the ability to deliver energy on demand; hydropower and geothermal power plants give direct access to renewable sources, while co-generation and other energy sources can use fuel from renewable sources (for example, hydrogen or biomass).

The challenge in designing a highly renewable electricity supply system (up to 100 per cent renewable) is to find the combination where advantages of each renewable source add up to a functioning and reliable system, while disadvantages are balanced out. Especially in the electrical system the need for reserve capacities, necessary as a back up for fluctuating sources, can be minimized by choosing the right combination of renewable technologies to minimize fluctuations. Demand management can also be introduced to get a better alignment between production and demand.

In the study the authors only studied the dynamic behaviour of the electrical system in the 'Fast Exit' scenario. This was done without optimizing the electrical energy system. The simulation was done for four typical weeks (in spring, summer, autumn and winter), with typical weather of the year 2006 (Generalitat de Catalunya, 2006). The optimization of the supply system and the introduction of modern electrical grid management methods (for example, demand management) will be investigated in a later study by Eurosolar Spain.

Considering the four simulated weeks as representative for all the four seasons of the year, the supply system according to the Fast Exit scenario is capable of supplying all the electricity demand in Catalonia. Generally solar and wind performance levels are substantially higher during spring and summer than they are in autumn and winter. Due to the strong spring and summer performance of fluctuating suppliers (solar and wind), it is often the case that photovoltaics, solar thermal power and wind energy can supply far more than the total electricity demand.

During the winter, the adjustable suppliers have a dominant role as a result of the decrease in solar radiation together with generally lower wind speeds. Looking at the big picture, climate variation over the year, with strong solar and wind performance during the warm periods, favours the system described here because the adjustable suppliers (hydropower, geothermal and biomass) have to contribute most during those times when they can be operated in the best way. While a high utilization of hydropower coincides with higher precipitation levels, geothermal and biomass plants can mainly be operated during times when there is a high demand for heat, thus giving the opportunity to take advantage of highly efficient CHP plants.

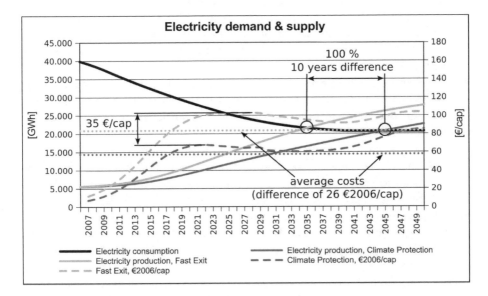

Figure 9.2 Development of electricity demand and two supply scenarios of Solar Catalonia

Source: Doleschek et al (2007)

General policy measures for going 100% renewable

The policy measures that the authors of Solar Catalonia propose in their study will perhaps help to change the situation. Economic, legal and institutional conditions for the energy system must fundamentally change and indeed this must happen soon. In practice, we will need to rely on a mixture of instruments and measures. In addition to what is described and planned, significant additional measures have to be taken to realize a sustainable energy future.

General political measures

Below is a list of suggested general political measures that would facilitate the path towards 100 per cent renewable energy supply:

- Adopt a set of rights and responsibilities that guarantee the democratization of the energy systems (see below).
- Develop a land use plan for renewable energies, based on a realistic picture of renewable energy potentials.
- Establish preferential areas for wind energy, according to the potentials and locations in each developed scenario.
- Assess and restructure the use of coastal areas for offshore wind energy, focused on the best locations.
- Set up an energy supply regime that favours renewable technologies as the first option whenever a new plant should be built.
- Give primacy of co-generation over conventional thermal power plants, combined with biomass and geothermal use as the first choice.
- Give priority to using pumped hydro plants to support and compensate fluctuating suppliers.
- Establish long-term electricity price guarantees for newly erected renewable energy plants and a permanent review of FiTs for the different technologies in order to keep the installation stimulus on a sufficient and technologically diversified level.
- Start a 'green government' initiative in public buildings and public services, with the improvement of energy efficiency in public buildings, with the incorporation of local energy generation and the replacement of the car fleet by most efficient vehicles, with priority give to biofuels, etc.
- Adopt energy efficiency standards for all electrical goods, with priority given to light bulbs and appliances (for example, all electrical appliances must meet the energy efficiency of today's most efficient appliance after two years).
- Establish a programme for promoting the monitoring and visualization of energy consumption (domestic, services level) in a way that will make consumption visible to users and more understandable than the reading of meters.
- Introduce, without delay, education and training on renewable energies, facilitating the fastest way to introduce and expand renewable technologies with assured quality.
- Introduce financing, legal and fiscal mechanisms and regulations in order to facilitate the above measures and technology research.

Besides the general policy measures it is also necessary to initiate concrete programmes and commitments such as:

- Establish a micro-co-generation programme with ambitious targets.
- Establish a solar roof programme with ambitious targets.
- Arrange an annual green community competition regarding local renewable energy generation.

- Arrange an annual 'zero energy buildings' competition.
- Establish a wind energy programme based on small (less than 5MW) wind farms with ambitious targets.
- Establish specific commitments, goals and targets to use public buildings for solar energy production and to start immediately emblematic or 'lighthouse' projects in the roofs and facades of the public buildings.
- Directly address celebrities/prominent entities to act as models and champions for utilizing solar energy or renewable energies in general.
- Promote local energy self-sufficiency programmes prioritizing the combined use of renewable energy resources existing in the area.
- Promote the development of a network of agencies or local energy centres independent of public authorities and energy companies, but with their participation, in order to pass the information about renewable energy and energy efficiency to the population.
- Create equitable partnerships between rural zones and urban zones, given that many rural zones could have a surplus of renewable sources of energy.

R&D has created renewable and efficient energy technologies for a permanent energy supply. Together, the political community and industry must take measures to implement a solar strategy. The measures described above are feasible and make sense. The most important step is to start now, since every day that goes by without enforcing a solar strategy only increases and complicates the problem. Because energy consumption is increasing, money is still being invested in fossil fuel systems and the finding of ways to solve the problem of climate change is merely being postponed.

Energy and democracy

In current energy systems, the right of people to take decisions is not respected and energy decisions are taken without any involvement of the people affected. In order to democratize and help establish a decentralized or distributed energy system in ways that are efficient, safe, clean and renewable, it is important for a society to recognize a set of basic energy rights. These rights are:

- the right to know the origin of the energy one uses;
- the right to know the ecological and social effects of the manner in which energy is supplied to each final user of energy services;
- the right to capture the energy sources that exist in the place where one lives;
- the right to generate one's own energy;
- the right to fair access to power networks and grids;
- the right to introduce into power networks energy generated in-situ;
- the right to a fair remuneration for the energy introduced into networks.

These rights have to be matched by a set of basic responsibilities:

- the responsibility to find out information;
- the responsibility to ask for information;
- the responsibility to generate energy with the most efficient and clean generation technologies available;
- the responsibility to use the most efficient end-use technologies available;
- the responsibility to conserve and use the generated energy with common sense, avoiding any kind of waste;
- the responsibility to limit oneself in the use of any form of energy;
- the responsibility to exercise solidarity with those underprivileged societies that have no or limited access to a clean means of energy generation as well as its final use.

Guaranteeing these rights should be one of the tasks to which governments give absolute priority. Exercising these responsibilities should be considered the fundamental duty of the responsible persons who depend on the sun as the source of energy. By adapting lifestyles to the solar energy flows (both direct solar energy and its indirect forms), people will discover that fewer costs of every kind will have to be borne in order to sustain life and prosperity on planet Earth.

References

Alemany, J. (1978) 'Energía alternativa: El poder del viento', *Ajoblanco*, November, pp28–30

Alemany, J. and Puig, J. (1980) 'Tecnología alternativa en Catalunya', *Transición*, no 23, Año III, Julio–Agosto, pp84–85

Autores varios (1979) *Alternativas: Recursos, Tecnologías, Construcción, Hábitat, Sanidad, Alimentación, Agricultura, Energías*, Euskal Bidea, Pamplona

Bosquet, M. and Gorz, A. (1977) *Écologie et liberté*, Éditions Galilée, Paris

Coleridge, S. T. (1898) 'The Rime of the Ancient Mariner', in *STC's Poetical Works*, University of Virginia, Charlottesville, VA

Comissió Tècnica d'Energia (1980) *Jornades d'Energia Eòlica*, 28–29 de gener, Comissió Tècnica d'Energia, Barcelona

Corominas, J. and Puig, J. (1982) 'L'autonomia energètica de Catalunya: Una opció possible', *Ciència*, 02, pp32–42

CPE-AEIC (1980) 'La tecnología per a l'aprofitament de la força del vent', 9–13 de juny, CPE-AEIC, Barcelona

de Cisneros, P., Ll. (ed) (1982) *Jornades de Política Industrial i Energètica*, vol I and II, Associació i Collegi d'Enginyers Industrials de Catalunya, Edicions Sirocco, vol II, pp150–178

Departament d'Indústria i Energia de la Generalitat (1981) 'Estudis i propostes tècniques per al desenvolupament de la política tecnològica i energètica del Govern de la Generalitat, ponències i comunicacions', Dept. d'Indústria i Energia, Barcelona

Dickson, D. (1975) *Alternative Technology and the Politics of Technical Change*, W. Collins Sons, Glasgow

Doleschek, A., Lehmann, H., Peters, S. and Puig, J. (2007) *Solar Catalonia: A Pathway to a 100% Renewable Energy System for Catalonia*, Fundació Terra, Barcelona, www.energiasostenible.org/sec3.asp?id_link=27&id_up=23 (in Catalan) and www.isusi.de/publications.html (in English)

Greenpeace Spain (2007) 'Renovables 100%: Un sistema eléctrico renovable para la España peninsular y su viabilidad económica', Greenpeace Spain, Madrid, www.greenpeace.org/espana/reports/informes-renovables-100

Generalitat de Catalunya (2006) 'Dades Meteorologiques Ema Integrades a La Xemec', Generalitat de Catalunya, Departament de Medi Ambient i Habitatge, Barcelona

Generalitat de Catalunya (2008) 'Projecte de Decret regulador del procediment administratiu aplicable per a la implantació de parcs eòlics i installacions FV a Catalunya', Generalitat de Catalunya, Barcelona

Groupe de Bellevue (1978) 'Projet Alter: Etude d'un avenir energetique pour la France axé sur le potentiel renouvelable. Esquise d'un regime a long terme tout solaire,' Syros, Paris

Harper, P. and Boyle, G. (1976) *Radical Technology*, Wildwood House, London

Leach, G., Lewi, C., Romig, F., van Buren, A. and Foley, G. (1979) *A Low Energy Strategy for the United Kindom*, International Institute for Environment and Development, London

Lönnroth, M., Johansson, T. B. and Steen, P. (1980) *Solar Versus Nuclear: Choosing Energy Futures, A Report Prepared for the Swedish Secretariat for Future Studies*, Pergamon Press, Oxford

Lovins, A. B. (1977) *Soft Energy Paths: Towards a Durable Peace*, FOE/Ballinger, Cambridge, MA

Lovins, A. B. (1979) *La Alternativa Energética: 1ª parte – La Energía Nuclear es un Fuego de Paja, 2ª parte – La Trayectoria no Emprendida en la Estrategia Energética*, Miraguano Ediciones, Madrid

Meseguer, C. (1978) 'Emplazamiento y anteproyecto de una central eólica de mediana potencia', Proyecto fin de carrera, ETSEIB, Barcelona

Ministerio de Industria y Energía and CEOE (1981) 'Plan para el fomento y la investigación e innovación tecnológica', Ministerio de Industria y Energía and CEOE, Madrid

Morris, D. (1979) *Planning for Energy Self-Reliance: A Case Study of the District of Columbia*, Institute for Local Self-Reliance, Washington, DC

Puig, J. (1979) 'L'energia eòlica i el seu futur a Catalunya, Jornades Catalanes d'Enginyeria', Palau de Congressos, Montjuïc, Barcelona, 28, 29 and 30 May

Puig, J. (1982) 'El passat i el futur de l'energia eòlica a Catalunya. Una aportació a la quantificació de la força del vent i una proposta per a la reintroducció del seu aprofitament', PhD thesis, ETSEIB, Barcelona

Puig, J. (2008) 'Barcelona and the Power of Solar Ordinances: Political will, Capacity Building and People's Participation', in Droege. P. (Ed.) 2008. *Urban Energy Transition: From Fossil Fuels to Renewable Power*, Elsevier.

Puig, J. and Corominas, J. (1984) 'Wind Energy in Catalonia: An Assessment of the Wind Potential and the Development of a 15 kW WECS', EWEC '84, 22–26 October, Hamburg

Puig, J., Meseguer, C. and Cabré, M. (1982) *El Poder del Viento: Manual Práctico para Conocer y Aprovechar la Fuerza del Viento*, Ecotopía Ediciones, Barcelona

Schumacher, E. F. (1975) *Small is Beautiful*, Blond & Briggs, London

TARA (1977) 'Alfalfa', *Extra*, Barcelona, pp1–36

Taylor, V. (1979) *The Easy Path Energy Plan*, Union of Concerned Scientists, Cambridge, MA

Todd, R. W. (1979) *An Alternative Energy Strategy for the United Kingdom*, National Centre for Alternative Technology, Machynlleth, Wales

Windirections (1984) WINDirections: Newsletter of the British and European Wind Energy Associations, vol IV, no 1, July

WorldWind (1984) 'Government-funded small wind turbine race rigged', *WorldWind*, September, p10

Chapter Ten

100% Renewable Transport

Andrew Went, Peter Newman and Wal James

In light of the mounting oil supply problem, the need to diversify the fuel sources of our transportation industry is growing increasingly urgent. The most promising, scalable, near-term solution is to switch to electric propulsion. The emphasis of our work so far has been on electrified public transport as the basis for transit-oriented development that can help transform cities into a less car-dependent form (Newman and Kenworthy, 1999; Newman et al, 2008; Went et al, 2008). The electric transit systems can then be switched to run on renewables and in some parts of the world this is being done: for example, Calgary's light rail runs on wind power. Even if this is dramatically successful, cities will still have cars and hence these, too, need to be switched to a renewable source of power. This chapter looks at how a new approach to electric vehicles can enable cities to not only have renewably powered private transport but how this can enable the city to become 100 per cent renewable in all its fuel and power needs.

Electricity offers several key advantages over other alternative fuel sources such as biofuels, hydrogen fuel cells, compressed air and CNG/liquefied petroleum gas (LPG) (Jamison Group, 2008). It is already ubiquitously accessible, requires no significant technological breakthrough, does not compete with land that can be used to grow food and can reduce mechanical complexity in the vehicle. Furthermore, urban air quality is greatly improved as electric propulsion emits no emissions from the vehicle's tailpipe and, if the electricity is supplied from renewable sources, no overall CO_2 emissions at all.

A new paradigm has arisen whereby the use of electric vehicles not only solves these oil and emissions issues, it also provides the chance to bring renewable energy into cities at a much larger scale than was possible before. Thus there is the potential for electric vehicles and renewable energy to enable cities to be carbon free. This combination of technologies is known as renewable transport.

Vehicle manufacturers have started to respond to consumer demand for more efficient vehicles by introducing hybrid electric vehicles (HEVs) and battery electric vehicles (BEVs). An intermediate between these is called a plug-in hybrid electric vehicle (PHEV), whose battery can be charged from grid electricity, allowing significant all-electric driving range with an internal combustion engine (ICE) available for longer distances. The high price premium of current battery technology and the fact that the majority of vehicles are driven for relatively short distances each day with the occasional longer trip, suggests that, in the near term, PHEVs will see the most rapid growth in these new vehicle technologies.

When a number of these PHEVs are parked and plugged in, their batteries represent a significant amount of stored energy that could potentially be fed back into the grid, hence the term 'vehicle to grid' or V2G. A large penetration of variable renewable energy (for example, wind and solar) in the grid will require an increase in ancillary services as well as storage support to avoid expensive fast-response fossil fuel backup generators when the renewable supply drops. The batteries in PHEVs can provide these services almost instantaneously upon request at low cost to the utility, and create a source of income for the vehicle owners that can help to offset the price premium and wear and tear on the vehicles. There will also be times when excess renewable energy is available that the batteries can absorb rather than being wasted. Such a system will require the use of sophisticated communication and control systems, known as a smart grid, that can allow the two way flow of electricity in a highly distributed system.

Hybrid technology: Green and gold

Civilizing the car has been an ongoing project in which many options have been pursued. The use of hybrid technology combines the best of green goals in terms of emissions and the best of commercial and consumer goals in terms of power and performance.

A hybrid electric vehicle is one that uses both electricity and a combustible fuel for propulsion. It is inherently more fuel efficient, to the order of over 30 per cent, compared to a similarly sized combustion only vehicle (DOT, 2008; Consumer Reports, 2009). To achieve these increases, hybrids take advantage of the complementary power generating characteristics of electric and ICEs at different speeds. They utilize the maximal power of electric motors at rest, which compensates for the corresponding minimum power available from the ICE at rest. This enables the ICE to be switched off at low speeds, which is where fuel consumption as well as all harmful emissions are at their relatively highest levels. The engine in a HEV generally does not start until the vehicle has reached a speed of 25 to 50km per hour.

Electrical assistance removes the constraint that the engine must match the instantaneous power demands of the driver, enabling engines to be designed for average power requirements, rather than the maximum. This allows much lighter and more optimized engine designs since they are not exclusively required to cope with extremes of operating conditions, i.e. low and high speeds and acceleration. They are optimized to perform most efficiently within a narrow operating range and are able to be kept in this position for longer due to the electric assistance.

Another efficiency enhancing measure is regenerative braking. It is the reverse of the process of providing electricity to the electric motors in the wheels, causing them to spin. When the brakes are applied the spinning wheels turn the electric motor, which generates electricity and charges the battery, slowing the vehicle in the process.

There are two main types of hybrids: parallel and series. In parallel HEVs, both mechanical and electrical energy power a mechanical transmission. The Toyota Prius is an example of a parallel HEV. In contrast, in a series HEV, only the electric motor powers the wheels directly. The engine is used to generate electricity for the motor and to recharge the battery. The lack of a mechanical link between the combustion engine and the wheels simplifies the overall design by eliminating the need for the conventional mechanical transmission elements (gearbox, transmission shafts and differential). The GM Volt series, due for mass production in 2010, will be a series of PHEVs.

The electric-only mode of series hybrids make them the preferred choice for most city driving, but for driving long distances at speed, they lose some ground due to the better efficiency of a mechanical transmission (~98 per cent) over an electrical transmission at high speeds (~70 per cent efficient). There are combined hybrid drive trains in development that can switch between series and parallel mode to optimize their transmission for any driving condition (General Motors, 2007).

Batteries

The main impediment to the widespread adoption of electric vehicles lies in their batteries, as all other components in the vehicles involve essentially mature technologies. Intensive R&D efforts have been mounted over the past 20 years to find this Holy Grail. The search now seems to be over. A new generation of batteries based around an advanced Lithium-ion chemistry appears to supply the energy and power densities required for an electric vehicle battery to provide sufficient driving range without being too bulky or overly heavy. They also have highly efficient charge and discharge cycles (Inderscience Publishers, 2008; Kang and Ceder, 2009), can be charged quickly, and have lifetimes comparable to other components in a vehicle, in the order of 10–15 years (House and Ross, 2007; General Motors, 2008). Furthermore the new chemistry has resolved the critical safety issues raised with earlier Li-ion batteries (American Chemical Society, 2007). The lithium-ion batteries currently available are superior to the lead-acid or nickel-metal hydride batteries used in early electric vehicles in almost every measurable performance indicator, except that the cost of these batteries is also much higher.

The recent flourish of research interest and funding into battery technology, spurred by the push for electric vehicles and demand for more powerful consumer electronics, should lead to even further increases, with some significant improvements already being claimed (Stanford University, 2007; American Chemical Society, 2008a; 2008b). The CEO of Toyota recently stated 'We are moving from the era of the gas tank to the era of the battery pack' when discussing the next generation of vehicles using Li-ion batteries (Toyota, 2008). Now that the search for a battery suitable for electronic vehicle use appears to be over, the advances in manufacturing techniques and technological optimization along with economies of scale should see the price premium reduce considerably over the next few years.

Plug-in hybrids: The next big step for cars

PHEVs are the next step towards a lower emission and more fuel-efficient transportation system. They will have large batteries that can recharge by plugging into the electricity grid using a standard wall socket, or a more specialized charging station for rapid charging. This allows an all electric driving range sufficient for a majority of short trips. An ICE is included to extend the driving range beyond the battery storage capacity but need only be engaged when driving longer distances. The all electric driving range can lead to significant reductions in fuel consumption provided the battery size is selected according to the daily commuted distance travelled. There is a trade-off between increasing the battery for extended all electric driving range and the size, weight and cost of these larger batteries. PHEVs will also have all the other efficiency features of a regular hybrid.

The reduction in fuel consumption will depend on the particular driving pattern of each vehicle. In an urban environment a typical day's commute will involve driving to and from work and may also include running some errands or attending a social activity. An average daily driving distance of 30km is a reasonable value for a predominantly car-dependent city with a substantial suburban component and is assumed as the daily distance travelled in the calculations in this chapter. A 'utility factor' is used to define the fraction of the total kilometres travelled electrically. A PHEV30km, meaning an all-electric driving range of 30km, was selected and should give a utility factor of around 50 per cent for average driving habits.

Modelling of the fuel efficiency in a mid-sized passenger vehicle – when configured as a conventional internal combustion vehicle (ICV), a HEV and a PHEV – has been conducted (Graham, 2001), revealing some startling findings, especially if the full production cycle of the fuel is included, which is known as the 'well to wheel' (WTW) energy (Simpson, 2005) and takes into account the energy required to extract and refine the fuel. The electric energy from the grid-charged batteries can be expressed in litres of unleaded petrol (ULP) fuel equivalent.

The results of the study, illustrated in Figure 10.1, show that a HEV uses 30 per cent less, and a PHEV 65 per cent less fuel than the ICV, with a similar saving on fuel costs. The PHEV also uses electricity, but in terms of energy content, in litres of ULP fuel equivalent, is a small amount and is considerably cheaper per kilometre travelled. The expected correlation between fuel use and greenhouse gas emissions for the ICV and HEV are also clearly indicated, however, the emissions from the PHEV will depend on how its electricity is generated. If the electricity is sourced from a coal-fired power plant then the overall emission reductions compared to the ICV are modest, but if the electricity is from a clean renewable source then there will be no further contribution to emissions and the PHEV will emit around 65 per cent less greenhouse gases for the same distance travelled.

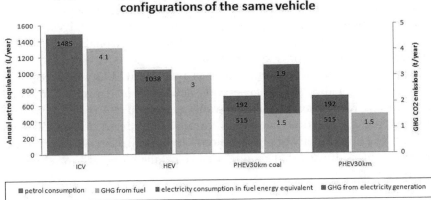

Comparison between fuel use and GHG emissions for several configurations of the same vehicle

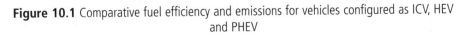

Figure 10.1 Comparative fuel efficiency and emissions for vehicles configured as ICV, HEV and PHEV

Source: Based on Graham (2001) and Simpson (2005)

Regardless of how the electricity is generated, it is important to note that the amount of greenhouse gases and harmful pollutants coming out of the PHEV's tailpipe are substantially reduced. Air pollution is a major problem in many large industrialized cities, leading to poor health and respiratory diseases. Even if the electricity was generated by a coal-fired power plant, the geographic relocation of the greenhouse gases from the tailpipe to the chimneystack of a power plant would lead to a huge improvement in air quality in the city itself. However, the greatest source of greenhouse gas emissions is grid electricity generation, primarily from coal-fired power plants. Any serious attempt at reducing total greenhouse gas emissions must confront this cause as well.

The significance of the emergence of plug-in vehicle technology – bikes, scooters, gophers and buses as well as cars – from an electric utility's perspective is that typically, the battery in an electric vehicle will be charged overnight when demand is low and when renewable energy such as wind is often discarded to avoid overloading the grid. Through this new market, utilities can sell electricity to allow renewables to be harnessed continuously and hence become commercially feasible, thereby enabling them to replace coal.

Renewables and the electricity supply industry: Finding a dancing partner

Renewable energy generation is seen by many as the saviour of the world's energy supply and climate change problems. Renewables such as wind, solar photovoltaic and solar thermal, hydro, geothermal and wave power offer the potential for virtually unlimited clean and cheap energy production. At current prices the cost of wind energy will approach parity with coal-fired power plants if the true cost of greenhouse gas emissions is factored in (Garnaut, 2008). However, fluctuating renewables such as wind and photovoltaics must include a storage system if they are to be a substantial part of the future in any city region. The potentially large storage supply in a fleet of PHEVs provides a means to enable this storage to occur, and at little additional cost since they are purchased primarily for transportation.

The particular mix of renewable options available will vary from region to region, however, a diversity of generation sources spread over a large area could allow a regular supply of renewable energy in much of the world's populated regions. Before renewables can be considered for supplying a significant fraction of a city or region's electricity supply there are several issues that need to be addressed that require a more detailed look at how electricity markets operate.

Electricity generators are divided into several categories based on the amount of electricity they can produce, how quickly they can change their output and the cost of the electricity. These are known as baseload, load following (or mid-merit), and peaking generators (see Figure 10.2). Baseload generators are large and hence expensive to build, but when running can produce large amounts of energy cheaply. The momentum of a large generator makes it difficult to change its output rapidly. Peaking generators are the cheapest to build but are more expensive to operate; this is one reason why they are only used during peak demand periods. Mid-merit generators have cost and performance characteristics between these two types of generators and increase output to cater to the normal ramping of demand as the day progresses.

Electricity demand is not constant throughout the day. It is generally related to the distribution of working hours throughout the day and is also temperature dependent. The heating and cooling of homes and workplaces, as well as of water, have a significant impact on the daily demand profile, with at least two distinct patterns. The summer profile typically peaks during the afternoon when it is hottest and more people turn their air conditioners on, while in winter two peaks occur, one related to heating workplaces in the morning and another when people get home and heat up their houses.

There are other components to the electricity supply system that add stability and generating capacity to the grid to ensure that supply and demand are balanced as much as possible. These are collectively known as frequency control ancillary services (FCAS) (Independent Market Operator, 2006).

Electricity networks must maintain 'spinning reserves' in case the demand is greater than anticipated or if a large generator experiences a forced outage. Typically, the larger a generator is, the longer it takes to warm up or increase its output, often taking so long for a large baseload generator that they are kept online continuously. However, the backup generators must be able to respond in a matter of seconds to prevent blackouts and so are kept spinning on a partial load, ready to be increased as needed.

Another ancillary service is called frequency control, also known as regulation. It makes possible the fine-tuned balancing of the supply with the demand. It is the most expensive service and must respond within seconds to ensure a constant voltage and frequency supply. The system must be able to supply small amounts of electricity to cover insufficient supply as well as being able to absorb any excess electricity to prevent a potentially destructive overload that could be caused, for example, by a large industrial consumer suddenly switching off. These ancillary services represent an additional expense on top of the cost of providing bulk power.

The amount of reserve capacity required is usually enough to meet the highest peak demand of the year while the largest generator is unavailable. Hence there is considerable oversupply of generating capacity in the system most of the time, especially at night. The shedding of power into cooling towers is thus necessary and is a heavy part of the needless production of greenhouse gases in this old-style power production.

Working with variability

The limits on the penetration of renewables into the grid arise from their variable output that requires an increase in ancillary services to ensure supply and demand can be matched continuously. Computer modelling based on weather forecasts can be used to predict the amount of electricity that can be produced from a renewable energy site a day or more in advance and this generation can be allocated to it by the network operator. There will be times that the renewable supply will exceed its expected output. Without storage, some of it would be discarded so as not to overload the system, since the baseload generators cannot adjust their output in the same timeframe as the renewable fluctuations. At the other extreme, if the renewable supply is less than that forecast, and if storage is included in the system, the stored energy can be used to delay or avoid the start up of fossil fuel backup generators.

While renewables such as solar and wind are variable in nature, they have the advantage of being predictably variable. It is true that the output from a single solar panel or wind turbine may vary considerably throughout the day, however, the statistical average of a large number of variable generators can be expected to follow a fairly smooth pattern, especially if they are spread out over a large area to reduce the impact that local effects, such as cloud coverage or wind lulls, would have on the group as a whole. The direct consequence of an increase in renewables is therefore a more stable supply since a larger number of generators would exist and they could be spread over a large area. Furthermore, with accurate forecasting, the amount of electricity that can be generated can be estimated fairly reliably. In Denmark, for example, where around 25 per cent of its electricity is supplied by wind farms, the 24-hour-ahead forecasts have a typical accuracy of within 7 per cent, while the four-hour-ahead forecasts are correct to within around 5 per cent (Morton et al, 2005).

It is unlikely that the generating capacity from a renewable source will drop suddenly and unexpectedly; it would typically be a more gradual and predictable decrease, as in the case of the sun setting or a pressure cell moving, causing wind speeds to change, on average, gradually. Should the largest generator in a network ever stop suddenly, several hundred megawatts of generating capacity would immediately be lost from the grid, however, even this loss is catered for by the requirements on the spinning reserves. If a single, or even several wind turbines were to suddenly cease functioning then only a few megawatts would be lost. With regard to the reserve capacity for renewables with storage, if the forecasting model is accurate then a smaller reserve margin could be maintained without lowering system reliability.

The increased take-up of PHEVs could supply the storage required for the reliable supply of renewables at low cost while also providing ancillary services. Since they would be able to feed energy back into the grid almost instantly upon request, they would be able to provide frequency control as well as spinning reserves, thereby eliminating much of the overhead operating costs associated with a large penetration of renewable energy into the grid. The increased storage could also allow the shifting of peak loads, thus 'flattening' demand.

The existing electric infrastructure is designed for a small number of large generators to power the grid. Output response is regulated by monitoring changes in the frequency of the AC voltage; if it falls slightly, output is increased and vice versa. A move to smaller, more numerous distributed generators of variable output that utilize a fleet of mobile PHEVs for ancillary services will require an upgrade in the infrastructure's communicating abilities.

Smart grids: The clever link

With the storage problem essentially solved, the introduction of renewables in large quantities still cannot occur unless an appropriate grid infrastructure, known as a smart grid, is provided. The most general definition of a smart grid is an electricity network that takes advantage of advanced two-way communications and computer-automated controls to streamline the provision of electricity services to consumers. A smart grid enables devices at all levels from utility to customer, to independently sense, anticipate and respond to real-time conditions by accessing, sharing and acting on real-time information. Thus the appropriation of all the benefits of renewable energy, and most other small-scale environmental technologies, requires this kind of smart control system – it is known as the ET-IT connection (Newman and Kenworthy 1999; Newman and Jennings, 2008).

Some of the benefits of a smart grid include: 1) increased operational efficiency by reduced distribution losses via optimal power factor performance and system balancing; 2) encouraging consumers to reduce their peak demand consumption by enabling time-of-use (TOU) rates and real-time consumption data; 3) allowing a wide variety of generation and storage options including small-scale renewable energy feed-in (and V2G); 4) a more resilient and self-healing network from fault anticipation and prevention; and 5) deferral of capital investment on distribution and transmission upgrades due to improved load estimates and reduction in peak load from enhanced demand side response (DSR).

In order to implement a smart grid, the first step is the introduction of smart meters. They not only communicate with utilities, but also extend into the home, providing information to consumers about their real-time consumption and can interact with 'smart appliances' also.

Governments and utilities worldwide are realizing the potential benefits of a more intelligent electricity network and are similarly investing in smarter rather than more powerful upgrades. Italy has 30 million; Canada and Sweden have each committed to 5 million installed by 2010; California will have smart meters for all residential locations by 2013; and Australia has also committed to a national rollout of smart meters, just to name a few (Ministerial Council on Energy, 2007). There are a range of options that vary in features and price but the design of these smart meters should incorporate the ability to be upgraded to streamline the adoption of expected advanced technologies such as smart grids and V2G.

A smart grid extends the communications from occurring between the end user and the utility to incorporating the substations, transmission networks and various electricity generators throughout the whole system, which involves more complex challenges. Already a substantially sized demonstration project is underway in Boulder, Colorado (Xcel Energy, 2008). The city was selected because it is a medium-sized metropolitan area (around 50,000 customers), is geographically concentrated and operationally well defined in terms of its electricity infrastructure. A number of options are being investigated to determine the optimum technical solutions. The project also includes integration of distributed renewable energy and a small-scale V2G demonstration (Parks et al, 2007).

Not only will a smart grid improve the operational efficiency, with the associated environmental benefits of reduced greenhouse gas emissions from polluting power stations, it also will provide the essential supporting infrastructure for the myriad of two-way communications required for the development of a fully integrated V2G network, or any large-scale deployment of distributed generation infrastructure, that will enable a significant amount of renewable energy sources to power the grid, cleanly and reliably.

V2G: The missing link

Smart grids and renewables linked to electric vehicle storage require the system of V2G to draw them together. The general premise of V2G is that a fleet of plugged-in vehicles represents a potentially significant resource that can be utilized to allow the introduction of a substantial proportion of renewable energy into the grid by acting as a buffer to regulate the variable renewable supply.

Although the generating capacity of an individual vehicle's engine is tiny compared to that of a large power plant, their combined numbers makes the combined capacity of a national fleet easily exceed that of its power plants. Similarly, a fleet of PHEVs will have a significant collective storage capacity in their grid-connected batteries.

Passenger vehicles are parked and perform no useful function for the majority of the day. With the advent of PHEVs, either hybrid or all electric, there is the prospect that idle vehicles can become assets that create value to their owners while parked (Brooks and Gage, 2001).

The introduction of V2G can be done in three stages, introduced conceptually here as 'supporting, shifting and stabilizing', and are distinguished by the number of vehicles available and the services they can provide. The adoption of V2G by PHEV owners is naturally enhanced by the early stages being more profitable, which helps to offset the price premium inherent in early adoption of any new technology.

The first stage, supporting, involves the use of PHEVs to provide ancillary services to the grid such as spinning reserves and frequency control. The batteries and power electronics in PHEVs are particularly well suited towards frequency control since the short, sudden demands on the battery are similar to the demands of driving. The net energy transfer of frequency control is typically zero, so only shallow cycling with minimal battery wear and no loss of storage result from providing this service. Spinning reserve payments would be made whenever a participating vehicle is plugged in, just for being available and, should they actually be called upon, an additional price is paid for any energy transfers. Spinning reserves could typically be required 20 times per year for 20 minutes each, while regulation could occur 400 times per day with small energy throughputs (Kempton and Tomic, 2005a).

Table 10.1 displays a comparison between the added initial cost of hybridizing a vehicle, with and without plug-in capability, with the savings in operating costs and revenue generated, as compared to a similarly sized ICV. The calculations put the price of unleaded petrol at AU$1.50 per litre ($4.26 per gallon) and assumes the price will remain at this value. Any price increase will effectively shorten the payback period. Table 10.1 shows that the payback period, for a PHEV providing ancillary services can be as little as just under two and a half years, even when the cost of extra wear and tear on the batteries is included (Brooks, 2002). Significantly, although the payback periods for a HEV and a PHEV not providing ancillary services are similar, the latter uses only half the fuel and emits only half the greenhouse gases.

Table 10.1 Payback periods for HEVs and PHEVs compared to standard ICVs

Vehicle type	Hybridization cost AU$	Fuel cost AU$/year	Maintenance AU$/year	Total F&M AU$/year	Savings AU$/year	Payback (years)	Payback V2G Freq. Con. (years)	Payback V2G Sp. Res. (years)
ICV	N/A	2224.8	300	2525	N/A	N/A	N/A	N/A
HEV	4727	1555.2	250	1805	720	6.6	N/A	N/A
PHEV	8457	182e + 777.6f = 959.6	200	1160	1365	6.2	2.3	4.7

Note: e=electricity cost; f = fuel cost (ULP).

The second stage, shifting, occurs when a critical mass of vehicles have saturated the ancillary services market, at which point the growing PHEV fleet will begin to have a combined storage large enough to allow excess energy generated during off-peak times (and prices) to supply the grid when it is under strain during peak demand periods, when the cost of generating electricity is much higher. This method of storing and shifting the energy supply is known as 'peak shaving'.

This is particularly relevant in the case of solar photovoltaic electricity since the solar intensity peak typically precedes the demand peak by only a few hours. The availability of a large amount of storage makes it possible to reconcile this time difference in peaks and makes the argument for solar power more attractive economically, especially if the value of the electricity generated from it is compared to the cost of the peak electricity it displaces.

A typical scenario could involve a person driving to work in all-electric mode, then plugging in to take advantage of the photovoltaic generation from a solar panel on their office's roof to recharge their batteries. This would also allow income to be generated by supplying ancillary services and, should the grid request it, automatically negotiate a rate above the minimum selling price set by the owner. Energy would then be discharged from the battery to a predetermined level of charge set by the owner in order to be able to drive home in all-electric mode.

Even if an owner does not have access to a solar panel, she or he can still profit by selling energy back to the grid. It is assumed that TOU price signals will compel most PHEV owners to charge up overnight using cheap off-peak electricity that can then be sold back to the grid later in the day at a higher price, to the benefit of both the vehicle owner and the utility.

Using some of the stored energy in PHEVs during peak demand periods will allow the utility to delay expensive additions to their peak power capabilities that are otherwise so underused they represent a burden on the total supply system that ultimately results in higher electricity prices. They will similarly be able to avoid expensive upgrades to their transmission infrastructure (transmission wires and substations) by meeting some peak demand locally, thereby not adding to the strain on the network and reducing transmission losses.

Furthermore, the use of electricity for private vehicle transportation presents a new source of income to the utility that need not require significant upgrades to their network if suitable incentives are in place to encourage off-peak charging, known as 'valley filling'. Because large coal-fired generators take so long to warm up or change speed, at night when demand is low they are often kept running faster than necessary just to ensure they will be able to get back to full speed when people start using electricity the next day, and so there is a lot of energy discarded as heat to the atmosphere and large amounts of greenhouse gases are needlessly emitted. Consequently, directing off-peak electricity to a useful purpose like recharging vehicles will reduce the amount of wasted energy.

This combination of peak shaving and valley filling results in the daily load profile having 'higher lows' and 'lower highs' and this results in a 'load levelling' effect that allows electricity generators to operate closer to their optimum conditions more of the time in a manner analogous to the efficiency increasing measures achieved by HEVs.

The ultimate manifestation of V2G, stabilizing, occurs when PHEVs occupy a large share of the vehicle fleet; their combined storage capacity being large enough to buffer the variable supply from renewable energy sources providing baseload energy. As more vehicles become available the services they can provide move into a less profitable domain, however, by this time the price premium on PHEVs should be reduced until they are likely to be only marginally more than a comparable ICE-only vehicle, and the fuel savings alone will still be a compelling reason to invest in a PHEV.

Wind power is currently the most competitive and widely scalable mature renewable energy source (CSIRO, 2008), when compared to geothermal, biomass or hydro. Calculations using the method proposed by Kempton and Tomic (2005b) demonstrate the number of plugged-in vehicles required to support a given penetration of wind generation providing baseload electricity, as shown in Table 10.2. Since not all the PHEVs will be plugged in or fully charged at any given time, the actual number of vehicles required will be higher. The capacity factor (CF) of any electricity generator is defined as the percentage of energy generated over a period of time compared to the maximum possible energy that could have been generated. For a wind turbine, provided it is in a suitable location, the CF can easily reach over 30 per cent (Sustainable Energy Australia, 2004). In comparison, the CF of a typical coal-fired power station is usually only around 60 per cent (ABARE, 2008). This highlights the underutilization required by the reserve margins as a consequence of there not being any storage in the grid.

The amount of wind power requested by the network operator can be tailored to match the forecast output based on weather conditions. Wind speed data modelling performed by Milligan (2001) on the operating reserves needed for high penetrations of wind, assuming geographically dispersed sites, have shown that a reserve margin of 15 per cent of the contracted generation, available for a full three hours is sufficient to cater for fluctuations from forecast output. Should a long wind lull be forecast then the utility can make alternative generating arrangements. Additionally, since this reserve is a battery (as opposed to a generator) it will not only cover shortfalls but also recover excess output over what is contracted from wind farms that otherwise would be lost.

Table 10.2 Wind capacity increases made possible, as a function of the number of PHEVs

Number	Vehicle of vehicles (GWh)	Vehicle storage available (GW)	Reserve capacity available for three hours (GW)	Max. wind installed name-plate (GW)	Average wind output @ 30 per cent CF (GW)	% Australia's total generation (29 GW).	Greenhouse gas reduction (Mt/yr)
1	12kWh	4.4 kW	1.46kW	9.75kW	2.92kW	N/A	N/A
1 million	12	4.4	1.46	9.75	2.92	10	26
2 million	24	8.8	2.92	19.46	5.84	20	52
3 million	36	13.2	4.38	29.19	8.76	30	79
4 million	48	17.6	5.84	38.92	11.68	40	105
5 million	60	22	7.3	48.6	14.5	50	131

Table 10.2 indicates that one million PHEV30kms would provide sufficient reserve capacity to enable an additional 2.92GW of wind average generation to the grid, which in the case of Australia would account for 10 per cent of its total generation. This does not include the approximately 20 per cent of wind penetration that could be currently supported by the existing reserve margins, as has been shown by studies into the UK and German electricity networks (Ilex Consulting, 2002; DENA, 2005).

Figure 10.2 illustrates a typical daily load profile for a possible future generation scenario in a fully realized V2G network. The generating portfolio consists of a large amount of renewable energy sources: wind power and others, and unspecified renewable sources as dependant on each sites' particular situation. A fleet of PHEVs is used to provide ancillary services, load levelling and reserve capacity for the large penetration of renewable energy in the grid.

For the vehicles considered earlier, each PHEV will emit around 2.6t/year of CO_2 less than its equivalent ICV model if the electricity is sourced from renewables, and hence 1 million PHEVs will emit 2.6 million tons/year less. If this million PHEVs was used to support 2.92GW of wind average generation then the greenhouse gas emission reductions, compared to generating 2.92GW from a coal-fired power station, would be around 26 million tons/year. These figures reveal that the greenhouse gas mitigation effect of V2G could be a factor of ten times greater in the stationary power generation industry than in the transportation industry.

To give renewables the best chance of being able to meet demand at all times, a diverse portfolio of renewable generators, spread over a large

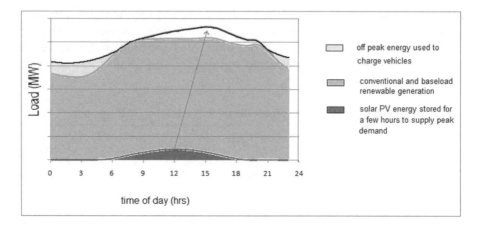

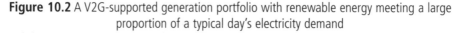

Figure 10.2 A V2G-supported generation portfolio with renewable energy meeting a large proportion of a typical day's electricity demand

Note: Note the distinction between baseload, load following and peaking types of generation as indicated by the load output on the y axis

geographical area, should be established that is tailored to optimize the particular renewable resources for each location. In theory, given enough storage and sufficient renewable resources, there is no limit to the level of penetration of renewables into the grid, up to and including 100 per cent.

Pathways to development: Taking the first steps with renewable transport

The first step in realizing the goals of renewable transport is transforming the vehicles themselves. The rate of uptake of PHEVs entering the market will depend on a number of factors, including: the price of petrol; the cost of batteries; the ability to scale up vehicle production; the availability of plug-in infrastructure in parking areas; the prices paid to vehicle owners for providing ancillary services; any financial incentive schemes by governments or otherwise that will lower the price premium and accelerate their market share, which could include tax cuts on low emission vehicles as well as encouragement for the establishment of a retrofitting industry. Carmakers worldwide are currently working frantically to develop and become market leaders in this new industry of plug-in vehicles (Calcars.org, 2008).

Many electric utilities are beginning to acknowledge the need to adapt to a new operating environment that takes into account the effect that greenhouse gas emissions, from the burning of fossil fuels for electricity generation, have on climate change. There is an urgent need to invest in electricity infrastructure around the world, as many existing networks were built 40–50 years ago and are now struggling to keep up with demand. These investment decisions must have a strategic and sustainable long-term view and must not be constrained by outdated definitions of energy infrastructure. Consequently, investment priorities are now tending towards developing smart solutions to facilitate increased energy efficiency and to develop on a significant scale embedded and distributed renewable energy systems that capitalize on each region's natural advantages with respect to renewable energy sources.

A number of interesting opportunities arise from the electrification of the vehicle fleet and subsequent relegation of oil-based fuels from the primary source of energy for transportation to playing a lesser role in which the fuel tank is regarded as a backup that may only be required occasionally. In this case, fuels with a lower energy density such as LPG or CNG, with a shorter driving range per tank of fuel, will appear as much less of a hindrance as they may seem to be today. These fuels are cheaper and cleaner per kilometre travelled compared to oil-based fuels and their utilization on a wider scale may provide some insulation from volatile international oil markets. PHEVs using LPG or CNG are expected to appear some time after petrol or diesel models however, as they will require more extensive infrastructure to match the availability of these fuel sources.

The introduction of PHEVs into the market is expected to be slow at first due to manufacturing-side limitations. A retrofitting industry that converts the legacy left from decades of ICE vehicle production, and more recent HEVs, into PHEVs can accelerate their adoption and also create many new jobs and business opportunities. As they will contain all the necessary power conditioning components required to output electricity at the appropriate voltage and frequency, PHEVs may also be attractive as uninterrupted power supplies for businesses or as mobile emergency electricity generators for essential services such as lighting and refrigeration to prevent food and medicines from spoiling.

The separate improvements of renewable transport, i.e. electric vehicle propulsion, renewable energy technologies and smart grids along with their greater combined effect due to V2G, provide the greatest opportunity for new and existing developments to drastically reduce their carbon footprint. These are the kind of dramatic synergies that can lead to exponential decline in the amount of greenhouse gases being emitted – something that the world must do if we are to avert rampant climate change.

To mainstream renewable transport, what is now required is a business case with pathways and gateways for the introduction of PHEVs, smart grids, V2G and the associated infrastructure and industry development as well as detailed investigations into the policy implications of developing this next stage of industrial innovation.

Policy implications of renewable transport

Renewable transport will not solve everything in the urban sustainability agenda

The breakthrough in technology we have called renewable transport solves the problems of fuel security and air emissions both from smog and greenhouse gases. It does not solve the social and economic problem of cities becoming too car dependent nor the issue of traffic congestion. These problems can only be solved by a better balance of transit (especially electric rail powered by renewables) and pedestrian and cycling infrastructure as well as urban design that can facilitate these modes such as transit-oriented development, corridor development, polycentric urban forms, walkable street designs and sustainable mobility management.

Renewable transport will require new policy partnerships

The partnerships required will be between:

- electric utilities, motor vehicle manufacturers and software companies to generate the seamless technology linking transport to the two-way storage and production of power from renewable energy;
- these technology providers and the land development industry who will now need to incorporate it in their designs;
- public authority planning agencies who will need to regulate the provision of infrastructure in homes, businesses and public spaces.

Renewable transport will require common good regulation

A whole series of questions will need to be debated and answered as part of the renewable transport transition. These include:

- What standards will be required on the provision of plug-in facilities in homes, businesses and public areas to ensure that there is consistency and not rival systems?
- How much of a city's public areas (streets where people park in particular) should be provided with plug-in facilities and how much should this be left to private parking operators to provide?
- What is required in terms of a government framework to enable this new technology to be demonstrated and facilitated in an integrated land development package?
- How can reduction of car dependence be linked to the provision of these new facilities?
- What can be done to ensure areas of poverty are not left out while wealthy eco-enclaves based on renewable transport are fitted out for the new economy?
- How can renewable energy be introduced rapidly enough to make the most of these new storage capacity opportunities?
- Will governments need to just facilitate this or regulate for a high renewables provision as part of all new urban development in the era when renewable technology is expected to take over?

The role of government incentives
Should governments (state or federal) provide incentives for the take up of renewable technology (i.e. through rebates, reduction in stamp duty, parking

concessions etc.)? If so, will this be complementary with national or international carbon trading mechanisms? One of the arguments of ET scheme advocates will be that the market-based method is to 'achieve abatement at the lowest cost to the economy', and therefore no other supports should be required. But as has been shown in Europe, ET schemes are not the most cost and prize effective, or fastest way to achieve uptake of the most effective technologies. They are also notional 'market mechanisms' (albeit highly artificial ones), and hence should not claim exclusivity over public incentives such as tax credits. It is therefore recommended to apply at least substantial public revenue signals (reduced or zero import and other taxes) to express the civic benefits and reduced environmental and health costs over internal combustion-based vehicles. Governments have a role to play in ensuring standards and frameworks are in place, as well as managing the take up of renewable technology.

Questions on standards and regulations include:

- How should national standards for renewable technology be developed (i.e. the requirements for smart meters, charging points etc.)? The challenge will be to match and align standards between national transport councils *and* national design rules *and* utilities *and* vehicle manufacturers *and* software companies.
- Could early movers such as companies like Better Place effectively set the standards? Will this be the best outcome for utilities and government?
- Given that the vehicle manufacturers operate in a global market there should be a global standard to manage interoperability of various systems, while maintaining safety and efficiency criteria.
- Who should make the investment in smart grid technology – utilities or renewable transport providers, or a combination of both? How will this be managed?

Questions regarding managing the take up of renewable transport include:

- Should the take up of renewable transport occur in geographic clusters or should it be spread across cities?
- What impact will clusters of renewable technology have on the grid?
- What is the potential for pilot projects; how should these be managed (i.e. through government departments or through geographic clusters)?

References

ABARE (Australian Bureau of Agricultural and Resource Economics) (2008) *Key statistics for the Australian electricity industry*, Table 28, www.abareconomics.com/publications_html/energy/energy_08/energyAUS08.pdf

American Chemical Society (2007) *Toward Improving the Safety of Lithium-ion Batteries*, 19 December, American Chemical Society, www.sciencedaily.com/releases/2007/12/071217110106.htm

American Chemical Society (2008a) *Improved Polymers For Lithium Ion Batteries Pave The Way For Next Generation Of Electric And Hybrid Cars*, 20 February, American Chemical Society, www.sciencedaily.com/releases/2008/02/080218160545.htm

American Chemical Society (2008b) *New Electrodes May Provide Safer, More Powerful Lithium-ion (Li-ion) Batteries*, 27 February, American Chemical Society, www.sciencedaily.com/releases/2008/02/080225092402.htm

Brooks, A. (2002) *Vehicle to grid Demonstration Project: Grid Regulation Ancillary Service with a Battery Electric Vehicle*, AC Propulsion, San Dimas, CA

Brooks, A. and Gage, T. (2001) 'Integration of electric drive vehicles with the electric power grid- a new value stream', presented at the 18th International Electric Vehicle Symposium and Exhibition, Berlin, Germany, 20–24 October, www.acpropulsion.com/EVS18/ACP V2G EVS18.pdf

Calcars.org (2008) 'How Carmakers are Responding to the Plug-In Hybrid Opportunity' www.calcars.org/carmakers.html

Consumer Reports (2009) 'The most fuel-efficient cars', www.consumerreports.org>

CSIRO (2008) *Fuel for Thought – The Future of Transport Fuels: Challenges and Opportunities*, CSIRO Future Fuels Forum, Campbell ACT

DENA (2005) 'DENA Grid Study', German Energy Agency, Berlin

DOT (2008) 'Summary of Fuel Economy Performance', US Department of Transportation, www.nhtsa.dot.gov

Garnaut, R. (2008) *The Garnaut Climate Change Review – Final Report*, Cambridge University Press, Cambridge

General Motors (2007) '*Global Alliance for Hybrid Drive Development: Cooperation between BMW, DaimlerChrysler and General Motors*', 9 July, http://media.gm.com/

General Motors (2008) 'GM says Chevy Volt battery will have a 10 year, 150,000 mile warranty', 5 August, www.gm-volt.com

Graham, R. (2001) *Comparing the Benefits and Impacts of Hybrid Electric Vehicle Options*, EPRI, Palo Alto, CA

House, E. and Ross, F. (2007) 'How to build a battery that lasts longer than a car', Altairnano Inc, www.powermanagementdesignline.com/

Ilex Consulting (2002) *Quantifying the System Costs of Additional Renewables in 2020*, Report to the UK Department of Trade and Industry, London

Independent Market Operator (2006) *Wholesale Electricity Market Design Summary*, September, Independent Market Operator, Perth

Inderscience Publishers (2008) 'Sweet Nanotech Batteries: Nanotechnology Could Solve Lithium Battery Charging Problems, 10 April, Inderscience Publishers, www.sciencedaily.com/releases/2008/04/080410101128.htm

Jamison Group (2008) *A Roadmap for Alternative Fuels in Australia: Ending Our Dependence on Oil*, Report to NRMA Motoring and Services, Sydney

Kang, B. and Ceder, G. (2009) 'Battery materials for ultrafast charging and discharging', *Nature*, no 458, pp190–193

Kempton, W. and Tomic, J. (2005a) 'Vehicle to grid fundamentals: Calculating capacity and net revenue', *Journal of Power Sources*, vol 144, no 1, pp268–279

Kempton, W. and Tomic, J. (*2005b*) '*Vehicle to grid implementation: From stabilizing the grid to supporting large-scale renewable energy*', University of Delaware, Newark, DE

Milligan, M. R. (2001) 'A chronological reliability model to assess operating reserve allocation to wind power plants', paper presented at the 2001 European Wind Energy Conference, Document #NREL/CP-500-30490, Copenhagen, Denmark, 2–6 July, www.osti.gov/bridge

Ministerial Council on Energy (2007) *Smart Meters Information Paper: On the Development of an Implementation Plan for the Roll-out of Smart Meters*, Appendix 1, Ministerial Council on Energy, Canberra

Morton, A., Cowdry, S. and Stevens, D. (2005) 'South West Interconnected System (SWIS) Maximising the Penetration of Intermittent Generation in the SWIS', Prepared for the Office of Energy, Western Australia

Newman, P. W. G. and Jennings, I. (2008) *Cities as Sustainable Ecosystems: Principles and Practices*, Island Press, Washington, DC

Newman, P. W. G. and Kenworthy, J. R. (1999) *Sustainability and Cities*, Island Press, Washington, DC

Newman, P. W. G., Beatley, T. and Boyer, H. (2008) *Resilient Cities: Responding to Peak Oil and Climate Change*, Island Press, Washington, DC

Parks, K., Denholm, P. and Markel, T. (2007) 'Costs and Emissions Associated with Plug-In Hybrid Electric Vehicle Charging in the Xcel Energy Colorado Service Territory', Technical Report NREL/TP-640-41410, www.nrel.gov/docs/fy07osti/41410.pdf

Simpson, A. (2005) 'Full- cycle assessment of Alternative Fuels for Light-Duty road vehicles in Australia', University of Queensland, Australia

Stanford University (2007) *New Nanowire Battery Holds 10 Times the Charge Of Existing Ones*, Stanford University, Stanford, MA

Sustainable Energy Australia (2004) 'Wind farming and the Australian electricity system', Sustainable Energy Australia, Melbourne

Toyota (2008) 'Toyota CEO's Presentation to Australia-Japan Industry Association', Perth, 13 October

Went, A., Newman, P. W. G. and James, W. (2008) 'Renewable Transport: How Renewable Energy and Electric Vehicles using Vehicle to Grid technology can make Carbon Free Urban Development', CUSP Discussion Paper 2008/1, Curtin University

Xcel Energy (2008) 'Smart Grid City Design Plan', www.xcelenergy.com/smartgrid

Chapter Eleven

Better Place

Peter Droege

In January 2008, the Israeli government announced a major initiative to switch the country to electric cars, with the 'Better Place' concept and company as the implementation platform, both developed by the software entrepreneur Shai Agassi. Since then, Better Place has grown powerfully and gathered an extraordinary momentum. What explains this success?

The long story of the electric vehicle is circuitous and has until recently been beset by tragic setbacks. Whatever manoeuvres and incentive misallocations have contributed to keeping the technology from becoming mainstream, a major stumbling block has been the challenge of transcendence: of overcoming the problem of realizing a new system within, or in spite of, the framework of an old infrastructure. The Better Place concept approaches this issue head on. It is a deceptively simple way to deal with fossil fuel dependency: use existing technology and market forces to transcend the old petrol engine by linking the initiative into a comprehensive, well-thought-out systems framework. This pivotal switch promises enormous opportunities to the energy industry, telecommunications, software developers and transport finance, a change of potentially historic proportions. The Better Place story provides a pragmatic yet visionary example of the ability of a few powerful individuals to bring a bright idea to realization.

The change is so potent because it fundamentally shifts the very paradigm in which one of the last unreconstructed bastions of early 20th century technology and ownership models still operates: the worldwide automotive system is based on a gargantuan and precarious petroleum refining and distribution network feeding the global fleet of ICEs, powering close to a billion motor vehicles around the world. A Better Place is, to date, the most comprehensive business model, among the recently rising world of electric vehicle suppliers, V2G planners and distributed mobile storage developers. It is also uniquely innovative, and relies on focusing on the leasing of batteries and the sale of access distance, or pay-per-mile service contracts, instead of the sale of cars and petroleum.

The venture is the brainchild of Shai Agassi, the Israeli-American Silicon

Valley entrepreneur, who was in line to become co-chief executive of SAP, one of the world's largest software companies, after having sold his own start-up company to SAP. Better Place's inaugural project, its implementation of an Israel-wide presence, was not the sole achievement of a single thought leader but the coming together of a range of interests, security considerations, advice by leaders such as Bill Clinton and Shimon Peres, and the willingness to fund on a large scale, both at a corporate and a personal level. Better Place, a company based in California, is to build the infrastructure, which is to involve a national grid of charging points and battery exchange stations. A few hundred vehicles were planned to be on the road in 2009 with production evolving based on market conditions by 2011; Renault-Nissan is committed to manufacturing the vehicles. The country's largest holding company and global player, Israel Corporation, initially invested $100 million, and Vantage Point Venture Partners, Edgar Bronfman Sr, and James D. Wolfensohn jointly contributed another $200 million to fund Better Place.

In 1990 the California Air Resources Board required that 10 per cent of new cars be non-polluting by 2003. While the initiative did not succeed, it set the stage for one of the most progressive energy policy frameworks. Inspired by this relatively enlightened state policy background, Shai Agassi developed the idea as part of the World Economic Forum's Young Global Leaders forum. Here the Young Global Leaders were confronted with the challenge of 'how to make the world a better place by 2020'. Agassi decided to focus on ridding the world of fossil fuel vehicles, using only existing technology, the force of the market and local demand, and a global proliferation strategy that would begin in one country and quickly lead to adoption by others. The idea soon became supported by Israel president Shimon Peres, who encouraged Agassi personally, helped convince automaker Renault-Nissan to be a partner, and assisted in raising the seed support needed to develop the concept into a self-funding business.

The Israel concept is built upon a partnership-based revenue approach, combining public policy innovation at the national level (tax breaks) and a major car manufacturer, Renault-Nissan. Initially, a fleet of 50,000 vehicles, a national matrix of 500,000 electric charging stations and 200 battery exchange points are envisioned, to be ideally powered by solar thermal or photovoltaic fields, wind and other renewable energy sources.

Project champion Shimon Peres sees Better Place not only as a pilot for island applications, but also for larger industrialized countries. On 13 January 2008 the tax on electrically powered vehicles was lowered by 10 per cent to encourage consumers. To make the concept a mass reality, Agassi's idea is to equip parking spaces with battery chargers; with 4 million parking spots in Israel, most cars can be kept charged for most of the time. For longer drives swapping a battery is to be as easy as filling up the car with petrol. The plan is based on an innovative battery exchange system, eliminating vehicle downtime for in-vehicle battery charging as required by plug-in hybrids or other electric vehicles since today's batteries have only a range of 100 miles.

Figure 11.1 A Better Place fleet car

Source: Better Place, www.betterplace.com (2009)

Pilot consumers will have to purchase a Renault-Nissan vehicle and subscribe to a Better Place service; Better Place will own the batteries. The concept is described as akin to mobile phone subscriptions, whereby like the mobile handset, the automobile itself becomes a secondary aspect to the primary focus of the business: the future delivery of affordable electricity for transport.

Agassi assumes a battery's life is limited to 1500 recharge cycles, hence the cost of running an electric car would be about 7 cents/mile, less than one-third of the cost of driving a gasoline-powered car, using 2008 prices. To overcome the typical time lag problem in innovation in the car industry (car longevity can easily exceed ten years, making rapid conversion to a new fleet impossible) the proposal is to offer free or low-cost conversions by replacing combustion engines with electric motors.

Within a year of announcement the initial Israel Corporation had also begun a related joint venture with Chinese automaker Chery. Denmark soon joined the Better Place programme, offering substantial tax incentives: lowering taxes to zero, while gasoline powered cars attract a heavy levy. Australia has also been added to the short but growing list of countries, with investment commitments by Macquarie Bank and Australian natural gas and energy utility AGL (see also Chapter 10). Better Place also successfully showcased its work in Japan, and important partners in both Northern California and Hawaii signed up at the end of 2008. Barely a year after the company's launch in Israel, Better Place was in discussion with more than 25 countries around the world. Since Barack Obama's election as 44th president of the US, Shai Agassi's sights are set on helping to replace a good share of the US's 250 million passenger vehicles with electric cars. Meanwhile, competition in business models and technology rises inexorably, and formerly reticent automobile producers awaken to the electric challenge.

Research input: Robyn Polan

References

Better Place (2009) 'Better Place', www.betterplace.com

Kiviat, B. (2008) 'Israel looks to electric cars', *Times*, www.time.com/time/world/article/0,8599,1705518,00.html

Markoff, J. (2007) 'Reimagining the automobile industry by selling the electricity', *New York Times*, 29 October, www.nytimes.com/2007/10/29/technology/29agassi.html

Vlasic, B. (2009) 'Mapping a global plan for car charging stations', *New York Times*, 9 February, www.nytimes.com/2009/02/09/business/09electric.html?_r=2

Chapter Twelve

How to Grow Food in the 100% Renewable City: Building-Integrated Agriculture

Viraj Puri and Ted Caplow

Increasing global urbanization and population growth, together with constraints on energy supply driven by climate change and resource limits, have highlighted the need for more sustainable cities in recent decades. Pressures on the rural landscape are no less serious, including mounting burdens of water pollution, soil degradation and irrigation demand on a finite agricultural base. As a response to both sets of concerns, bringing agriculture directly into the built environment has the potential to reduce ecological impacts, cut fossil fuel-based transportation, enhance food security, save building energy and enrich the lives of building occupants.

Despite the absence of significant arable land in most cities, an opportunity exists to produce meaningful quantities of food for large urban populations, within the built environment. By combining technically sophisticated, commercially proven controlled environment agriculture (CEA) techniques with unique energy saving innovations, cities can feed themselves both efficiently and cost effectively, while sharply reducing energy use and environmental impact. CEA, also known as greenhouse agriculture, combines horticulture and engineering to optimize crop production, quality and yield. Hydroponics, the culture of plants in water, is particularly well suited for urban applications of greenhouse agriculture. By employing environmental design, hydroponic greenhouse methods can be adapted for use on and in buildings within cities; a concept known as building-integrated agriculture.

New York City alone has approximately 14,000 acres of unshaded rooftop space in its five boroughs (Center for Energy, Marine Transportation and Public Policy, 2006). Based on modest commercial hydroponic production yields (10 pounds per square foot average leaf and vine crop yields reported on the Science Barge, New York City) and the per capita fresh vegetable consumption in the US (173.2 pounds per capita, excluding potatoes; see http://postharvest.ucdavis.edu/datastorefiles/234-66.pdf), this unused rooftop space is capable of meeting the vegetable needs of over 30 million people – more than 3 times the population of the city. Building-integrated agriculture has the potential to eliminate the use of fossil fuels in the production and transportation of fresh fruit and vegetables while enhancing the livelihoods, health and nutrition of urban dwellers worldwide.

Background

Increasing urbanization and the global construction boom have underscored the importance of efficiency in the built environment. In the US, buildings account for 39 per cent of energy use, 68 per cent of electricity consumption, 12 per cent of water consumption and 38 per cent of CO_2 emissions (US EPA, 2004). Figures for Europe are similar (Balaras et al, 2007).

In a less tangible shift, the natural world is being marginalized and green space is increasingly remote as people live and work in ever-taller structures. Agriculture has an equally significant impact on our world. Modern agriculture feeds billions of mouths every day, but is the world's largest consumer of land and water, the source of most water pollution, and the source of an estimated 15 per cent of the world's greenhouse gas emissions (Netherlands Environmental Assessment Agency, 2005).

Three major trends will strain the global food system over the next half-century and place significant additional pressure on the environment. First, according to official UN estimates, global population is expected to exceed 9 billion by 2050. Second, more than two-thirds of these people are expected to be urban dwellers (Montgomery, 2008). Urbanization requires that food, once grown and harvested, must travel hundreds or thousands of kilometres to reach consumers. Fresh produce travels an average of 2500km to reach US cities, adding to fossil fuel consumption, traffic congestion, air pollution and carbon emissions (Pirog and Benjamin, 2003). Third, global warming is predicted to lead to widespread shortages of food, water and arable land by 2050 within a broad belt extending north and south of the equator and encompassing some of the world's most populous regions (Parry et al, 2007; UNEP, 2007; Brown and Funk, 2008; Lobell et al, 2008).

Today's notion of 'green building' does not appear to be green enough, nor widely enough applied, to reverse these trends. A more aggressive solution could be within reach. Growing food crops on buildings could reduce our environmental footprint, cut transportation costs, enhance food security, save energy within the building envelope, and enrich the physical and psychological comfort of building occupants and city dwellers.

Building integrated agriculture

Ecological sustainability requires minimizing the use of non-renewable resources while protecting other species and their habitat. Building-integrated agriculture is based on the idea of growing food in the built environment, using renewable, local sources of energy and water.

This concept is a response both to the increasing energy burden and to increasing energy price risk, which is particularly hard felt in the agricultural sector, including CEA. Building-integrated agriculture also addresses ecological and public health concerns surrounding conventional agriculture, including climate change, high resource consumption, long distance food transport, food security and food safety.

Hydroponics, the culture of plants in water, is a technically sophisticated means of agriculture practised in most regions of the world. Essential mineral nutrients are dissolved in the water used to irrigate the plants, eliminating the need for soil. Without the need for soil, greenhouse facilities become substantially more modular. Recirculating hydroponics, the most modern and environmentally sustainable method, can produce premium-quality vegetables and fruits using up to 20 times less land and 10 times less water than conventional agriculture, while eliminating chemical pesticides, fertilizer runoff, and carbon emissions from farm machinery and long-distance transport (Vogel, 2008).

Figure 12.1 Recirculating hydroponics: The most water-efficient form of agriculture in the world

Source: © BrightFarm Systems

Hydroponics is practised on industrial scales in The Netherlands, Israel, Spain, UK, Mexico, Canada and the US. Commercial greenhouses in Almeria, Spain, for example, where much of the salad crops consumed in the UK are grown, cover an area of nearly 20,000 hectares. Today, there are over 1200 hectares of hydroponic vegetables produced in the US, Canada and Mexico. In 2008, the Fresca Group began operating a new 91 hectare greenhouse facility on the Isle of Thanet, Kent.

Figure 12.2 Tomato crops under hydroponic cultivation

Source: © BrightFarm Systems

Crop quality and yield are largely a function of climate control within the greenhouse environment. Maintaining constant temperature and humidity levels in a greenhouse allows year-round crop production but also presents an energy and efficiency challenge. In northern latitudes, wintertime heating accounts for the majority of energy demand and CO_2 emissions; nearly all of this heating need is met using fossil fuels.

When compared with a conventional, slab-mounted greenhouse, rooftop integration yields direct energy savings by eliminating heating losses through the building roof and the greenhouse floor, and by capturing waste heat from the building exhaust air. Special greenhouse design features, including double glazing and a thermal blanket, can result in substantial additional reductions in heating demand. Locating the project in a dense urban area, where temperatures are warmer due to the urban heat island effect, also plays an important role in reducing heat demand. The remaining heating needs of the facility can be met using renewable fuels, such as biodiesel or waste vegetable oil, virtually eliminating net CO_2 emissions from heating.

In warmer climates, cooling loads present the energy challenge. A greenhouse placed on the roof of an urban building provides a suitable space to implement a large evaporative cooling system for the combined structure. Without the greenhouse, evaporative cooling systems would likely be unfeasible for the building due to constraints of space, humidity and/or cost. Energy can also be saved in the combined structure by the elimination of solar gain and thermal losses through the building roof, because this surface now becomes the floor of the greenhouse, with approximately the same temperature above and below.

The electrical needs of a building-integrated agriculture facility can be met by on-site solar photovoltaics. Solar photovoltaics are a particularly appropriate fit for CEA as peak electrical demand coincides with peak electrical supply: strong sunlight on a hot summer afternoon. Electrical load can be minimized by using natural ventilation, evaporative cooling and high-efficiency pumps and fans.

As a result of these energy saving strategies, building-integrated agriculture can produce vegetables with a lower total energy input, per kilogramme of delivered product than either conventional greenhouse agriculture or conventional field agriculture, and with a much lower carbon emissions profile. The environmental benefits are rounded out by land and water savings, and the elimination of fertilizer, pesticide and stormwater runoff in the highly impermeable urban landscape.

Increased urban green space, in the form of building-integrated agriculture, can also help mitigate the urban heat island effect: the phenomenon of city air temperatures rising up to 10°C higher than surrounding non-urban areas because of the abundance of dark, heat-absorbing surfaces such as rooftops and pavement. This effect increases the demand for air conditioning, which subsequently increases emissions. Heat islands also increase ground-level ozone, formed when heat, sunlight and air-borne chemicals mix, which worsens symptoms of asthma, emphysema and lung cancer.

Building-integrated agriculture can reduce this phenomenon effectively by reducing a building's cooling loads through evapo-transpiration, as well as reducing heating loads by adding mass and thermal resistance value to the building. Rooftop greenhouses can often be more effective than traditional green roofs, as the enclosed glazed area can act as a passive heat reservoir.

Building-integrated agriculture's proximity to the retail market sharply reduces transportation fuel consumption and associated air emissions, and subsequently the public health impacts related to exposure to particulate matter and smog. One of the highest rates of respiratory ailments in the world, including asthma, is found in the South Bronx area of New York City. The South Bronx is also home to the world's largest wholesale food market. This community contends with the emissions of 20,000 diesel truck trips into and out of the market each week. Trucks often idle for up to 12 hours at a stretch while waiting for goods or to comply with regulations.

Millions of city dwellers worldwide live in areas with limited opportunities to obtain fresh produce. These areas tend to have the highest levels of diet-related diseases: obesity and diabetes in developed countries, and anaemia, scurvy and rickets in the developing world. As health experts continue to recommend adding more fruit and vegetables to a healthy daily diet, it becomes increasingly important that consumers have access to safe, high-quality produce. Produce grown in sterile greenhouse conditions also reduces or eliminates the risk of pathogens. The proportion of food-borne illness associated with fresh fruit and vegetables has increased over the last several years. Food produced in the city, for local residents, need not travel more than a few kilometres, reducing the handling. This proximity to the end-user ensures not only a fresher, more nutritious product but also greater control over the food delivery chain.

As an added benefit, building-integrated agriculture can deliver much needed 'green collar' jobs to urban areas. Creating thousands of green collar jobs, such as building and operating greenhouse facilities, installing solar panels and constructing green buildings, is an excellent way to fight both global warming and urban poverty. The livelihoods of the disadvantaged and the chronically unemployed in many communities under climate change pressures can be stabilized or improved.

Horizontal rooftop greenhouses

Situating environmentally low-impact hydroponic greenhouses on new or existing flat roofs in cities, provides the urban farm with more space than would have been available at ground level. Hydroponic greenhouses are relatively light and thus installation on rooftops does not normally require significant structural reinforcement to the host building.

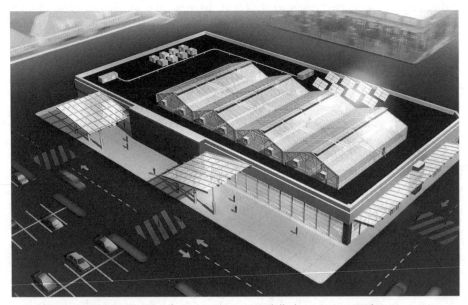

Figure 12.3 Rooftop greenhouse modelled on a supermarket

Source: © BrightFarm Systems

Applications for horizontal rooftop greenhouses range from commercial-scale farms on shopping malls, warehouses and other large buildings, to smaller systems designed for schools and community housing. At a scale of 1000m^2 and up, a rooftop greenhouse can offer commercial fresh produce strategically located at the point of consumption. Entrepreneurs can take advantage of resource efficient, high-yield, year-round production with reduced food distribution costs.

Building-integrated agriculture can provide fresh produce to residents of apartment buildings and community housing projects regardless of the location of the nearest supermarket. Greenhouse facilities can be leased out to a commercial operator, or operated as a community-supported agriculture (CSA) system.

Figure 12.4 Rooftop greenhouse modelled on a residential tower

Source: © BrightFarm Systems

School and university rooftops are excellent sites for rooftop greenhouses, yielding both nutritional and pedagogical benefits. Building-integrated agriculture offers hands-on educational opportunities in the basic sciences as well as environmental topics, and 'green' learning environments have been demonstrated to improve student focus and performance. Onsite production of nutritional vegetables provides healthy food for the student body, while simultaneously emphasizing food quality and origins in the curriculum.

Horizontal rooftop farms are pragmatic and require little new technology. The environmental benefits are significant: each hectare of rooftop vegetable farm, if built in the US, would, on average, free up 10 hectares of rural land, save 74,100 tons of fresh water each year, and if fully integrated with building heating systems and onsite solar power, eliminate 988t of CO_2 emissions per year compared with a conventional greenhouse.

The vertically integrated greenhouse

The vertically integrated greenhouse (VIG) is a concept for a highly productive, lightweight, modular, climatically responsive system for growing vegetables on a vertical curtain wall facade that was designed by an interdisciplinary team led by New York-based BrightFarm Systems, with contributions from the fields of ecological engineering, plant science, architecture and HVAC engineering.

In the building sector, the double skin facade (DSF) is an innovation that can reduce the energy used for space conditioning in modern high rise buildings by up to 30 per cent (Streicher, 2005; see also www.battlemccarthy. com). A DSF consists of a vertically continuous void space enclosed by a second curtain of glazing over the entire facade. A double skin provides solar heat in winter, buoyancy driven cooling flows in summer, and allows opening windows year round. Despite these advantages, DSF applications remain limited due to economic concerns and the need to install a large shading system within the cavity to realize the full benefits.

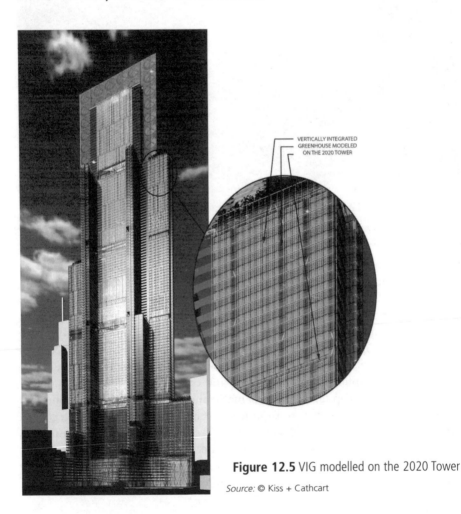

VERTICALLY INTEGRATED
GREENHOUSE MODELED
ON THE 2020 TOWER

Figure 12.5 VIG modelled on the 2020 Tower

Source: © Kiss + Cathcart

The VIG combines a DSF with a novel system of hydroponic food production for installation on new high-rise buildings and as a potential retrofit on existing buildings. In this design, crops are cultivated behind a glazed curtain wall on the southern facade of a building, on an array of horizontal trays suspended on vertical cables. The VIG functions alternatively as an adaptive solar energy capture device and a biological shading system, in winter and summer respectively. Hence, in addition to producing food, the installed plants, in effect, reduce building maintenance costs by providing shade, air treatment and evaporative cooling to building occupants (Stec, 2005).

Figure 12.6 Lettuce crops in a VIG office facade

Source: © Kiss + Cathcart, BrightFarm Systems

By adding commercial-scale vegetable production, the VIG aims to sufficiently strengthen the economic argument for the DSF, and thereby provide the leverage necessary for widespread adoption of double facade technology. At the same time, by meeting an ecologically significant share of the food demand of building occupants, the VIG advances a more holistic set of expectations for the performance of green buildings, which considers the resource consumption not only of the building itself but also of its occupants.

BrightFarm Systems have developed a working prototype. New steps in the development of the system include refining the design of an appropriate irrigation system, developing a more detailed computational modelling of

Figure 12.7 VIG prototype undergoing testing

Source: © BrightFarm Systems

heating and cooling processes, selecting optimal glazing materials, optimizing the structural design of the curtain wall and testing additional working prototypes.

Constraints and next steps

The growth of building-integrated agriculture faces barriers to widespread implementation, including appropriate sites, zoning and building regulations and permitting, efficient distribution models, as well as certain technical challenges, particularly with regard to robust, replicable energy saving innovations.

While small residential buildings or buildings with pitched roofs are less likely to be worth the capital outlay of these systems, a careful look at the urban landscape reveals a significant amount of appropriately sized flat roofs on supermarkets, warehouse roofs, school and hospital roofs or shopping centres.

Greenhouse agriculture does not address the efficient production of all agricultural products. Hydroponic techniques are best suited economically and logistically to a range of vegetables that include: leaf crops such as spinach, lettuce and other salad greens; vine crops such as tomato, cucumber, pepper, squash, beans, zucchini; and culinary herbs such as basil, parsley, chives and coriander. These perishable crops travel a long way from farm to table and need to be kept cold and fresh en route, increasing the energy burden. Most vegetables have extremely high water content; shipping dehydrated storable foods makes infinitely more sense than, for instance, tomatoes, with 90 per cent water weight, across thousands of kilometres. Grains are better suited to be grown in conventional field settings. Grains, however, can be stored for long periods of time and can be shipped dry, without the need for climate control.

Furthermore, fresh perishable vegetables tend to suffer the most in terms of taste and vitamin content from being transported long distances, or being harvested before they are ready in order to allow for transportation. Even by bringing fresh fruit, vegetables and herb production into cities, we can allow millions of hectares of agricultural landscape to return to diverse ecosystems, vital to the preservation of life on the planet.

Building-integrated agriculture is highly compatible with bioclimatic design principles, advanced by architects and environmental designers, such as Ken Yeang, over recent decades. Food crops grown on, and in, buildings can provide heating, cooling and ventilation – services for the building that decrease energy use while improving the microclimate around buildings. In addition to thermal comfort and energy savings, food crops can enrich the aesthetics and the psychological comfort of building inhabitants. Studies have indicated that a 'green' work environment raises worker productivity by 1.0 to 1.5 per cent (Kats, 2003).

In addition to existing cities, building-integrated agriculture has a role to play in new, planned urban developments, including 'ecocities'. Ecocity is increasingly being used to describe a development initiative aimed at minimizing both external inputs of fossil fuel energy, water and food and outputs of sewage, garbage, heat, pollution and CO_2. Ecocities currently under development include Dongtan in China and Masdar in the United Arab Emirates. Integrated, ecologically sound food production systems would form a central feature of the truly renewable ecocities of the future.

The past few years have seen the proliferation of green roofs and green walls created by designers, architects, urban gardeners and artists for their economical, environmental and aesthetic impact. From a more productive standpoint, rooftop farms and vertical farms are being envisioned as new ways to grow meaningful amounts of food for local populations. The widespread adoption of urban farms requires continued research, development and testing as well as the construction of working proofs of the concept.

Conclusion

Locating highly productive, environmentally robust food production systems not only in our cities but actually on buildings, enriches the lives of city dwellers, while offering a solution to some of the integral challenges facing the 100 per cent renewable city: reducing fossil fuel consumption; reducing the environmental impact of growing food; reducing the distance food travels before reaching urban consumers; and reducing the environmental impact of buildings.

Figure 12.8 The Science Barge: A prototype urban CEA facility built by BrightFarm Systems

Source: © BrightFarm Systems

References

Balaras, C. A., Gaglia, A., Georgopoulou, E., Mirasgedis, S., Sarafidis, Y. and Lalas, D. (2007) 'European residential buildings and empirical assessment of the Hellenic building stock, energy consumption, emissions and potential energy savings', *Building and Environment*, vol 42, no 3, pp1298–1314

Brown, M. E. and Funk, C. C. (2008) 'Food security under climate change', *Science*, no 319, p580

Center for Energy, Marine Transportation and Public Policy (2006) *Powering Forward: Incorporating Renewable Energy into New York City's Energy Future*, Graduate study of the Urban Energy Policy Workshop, School for International and Public Affairs, Columbia University, New York

Kats, G. (2003) *The Costs and Financial Benefits of Green Buildings: A Report to California's Sustainable Building Taskforce*, California Sustainable Buildings Task Force, Sacramento, CA

Lobell, D. B., Burke, M. B., Tebaldi, C., Mastrandrea, M. D., Falcon, W. P. and Naylor, R. L. (2008) 'Prioritizing climate change adaptation needs for food security in 2030', *Science*, no 319, p607

Montgomery, M. R. (2008) 'The urban transformation of the developing world', *Science*, no 319, pp761–764

Netherlands Environmental Assessment Agency (2005), *Emission Database for Global Atmospheric Research*, Netherlands Environmental Assessment Agency, Bilthoven

Parry, M. L., Canziani, O. F., Palutikof, J. P, van der Linden, P. J. and Hanson, C. E. (eds) (2007) 'Summary for Policy Makers', in *Climate Change 2007: Impacts, Adaptation, and Vulnerability*, Contribution of Working Group II to the Fourth Assessment Report of the Intergovernmental Panel on Climate Change, Cambridge University Press, Cambridge, pp11–13

Pirog, R. and Benjamin, A. (2003) *Checking the Food Odometer: Comparing Food Miles for Local Versus Conventional Produce Sales to Iowa Institutions*, Iowa State University, Ames

Stec, W. J. (2005) 'Modeling the double skin façade with plants', *Energy and Buildings*, vol 37, pp419–427

Streicher, W. (2005) 'BESTFAÇADE: Best Practices for Double Skin Façades', in *WP1 Report: 'State of the Art'*, European Commission, Brussels

UNEP (United Nations Environment Programme) (2007) *The Global Environmental Outlook 4: Environment for Development*, UNEP, Nairobi

US EPA (2004) *Buildings and the Environment: A Statistical Summary*, US EPA, Washington, DC

Vogel, G. (2008) 'Upending the traditional farm', *Science*, no,319, pp752–753

Chapter Thirteen

Masdar City Master Plan: The Design and Engineering Strategies

Matthias Schuler

Big Oil to Big Solar: A city for the Masdar Initiative

Preface by Peter Droege

Abu Dhabi's Masdar Initiative was launched in April 2008, a broad-based, multi-billion dollar investment strategy in renewable energy and 'clean technology' (http://www.masdaruae.com). As part of the initiative, the 'Masdar City' development is to be constructed 17km south east of Abu Dhabi's urban core, linked into the adjacent airport and other surrounding communities. It is planned as a new and independent urban center for a day- and night-time population of about 50,000, set within a larger precinct that is home to some 470,000 people. The development is pushing the 'sustainable business park' concept to a higher level of technological aspiration, theming it with local climate response, efficiency and renewable energy virtues in keeping with the scale that has become the hallmark of regional projects, before the global financial crisis of the late 2000s struck, and signs of overbuilding came to emerge in the wider region. While the socio-economic and cultural dimensions - and risks - of attempting a new town-within-the-city are by definition an open question, the Masdar Initiative's initial renewable city building ambitions were extraordinary, promising to go far beyond public relations.

Four goals motivated the plan's champions, the Government of Abu Dhabi and the Abu Dhabi Future Energy Company (ADFEC): to diversify Abu Dhabi's economy; expand the country's global energy market reach; nurture new renewable technologies; and to contribute to 'sustainable human development'. At the energy and economic end of the equation it is hoped that Masdar will help save US$2 billion worth of oil over the next 25 years, at the same time boosting Abu Dhabi's GDP by

2 per cent. Carbon finance is expected to be a prime investment source.

The ADFEC headquarters were key to the business strategy, as was the creation of the Masdar Institute of Science and Technology. This organization conducted global research to support Abu Dhabi's quest of controlling a significant, diversified and integrated segment of the global renewable energy industry. The physical development strategy of Masdar City, described by one of the key contributors to the development's design in the next chapter, relied on trademark buildings as core features: Adrian Smith and Gordon Gill's solar powered Masdar Headquarters building is hoped to generate more electricity than it consumes. Initially budgeted at US$300 million, the 120,000m2 structure is to anchor the centre of the development. As a mixed-use initiative it was conceived to house the Masdar offices, private residences and a businesses support facility. ADFEC and WWF engaged in an eco-promotional partnership in which Masdar City is being assessed using the principles and objectives of WWF's 'One Planet Living' Programme, developed by WWF and the UK's BioRegional group. In January 2008, WWF and ADFEC presented the 'Sustainability Action Plan' at Abu Dhabi's first World Future Energy Summit.

The physical design concept is led by a design team headed by Foster and Partners, involving transport planners, climate, renewable energy, efficiency and infrastructure systems experts. Masdar City received the first World Clean Energy Award from the Transatlantic 21 Association and was named 'Sustainable Region/City of the Year' at Euromoney and Ernst & Young's 2008 Global Renewable Energy Awards. As one typical example of the kind of ventures to be spawned in the complex, a joint project on hydrogen development between BP Alternative Energy and Rio Tinto was announced in 2008.

The project's model city aspirations are pitted against the stark reality of the global petrochemical legacy. The surroundings feature a Formula 1 race track, a car oriented shopping mall and an aluminium smelter. The wider Abu Dhabi urban system may receive positive restructuring signals from Masdar City, as the attention inevitably will need to shift to developing, infilling, refurbishing and enhancing existing areas in equally visionary ways. The potential is clear: the time has come to convert the anachronistic Formula 1 into Formula E - using electric vehicles, and the emirate does have the capability of converting its abundant solar reserves into say, global renewable hydrogen powered eco-industrial production capacity.

The nomination of Masdar and its headquarter as central seat of the International Renewable Energy Agency (IRENA) in 2009 further brightens the spotlight on a promising beginning even further. The Masdar initiative's city precinct project shines brightly as the model for all new development and building it hopes to become anywhere.

Research input: Robyn Polan

Introduction

A team made up of architects and transport planners, as well as climate, infrastructure and renewable energy systems engineers, developed an integrated design approach to sustainable urban planning. Foster and Partners designed the 6.4km² (640 hectare) precinct, which should eventually provide living space for around 50,000 inhabitants, based on the principles of the WWF 'One Planet Living Sustainability Standard'. This standard also includes the prescribed targets for the city's 'ecological footprint'. Masdar City will even surpass the requirements of the ten sustainability principles: zero carbon, zero waste, sustainable transport, sustainable materials, sustainable food, sustainable water, habitats and wildlife, culture and heritage, equity and fair trade, health and happiness. It is envisaged that the Masdar Institute of Science and Technology will become a leading R&D centre for the application and utilization of renewable energy sources.

Climate and culture: Reflecting the local conditions

Within the international design team, Transsolar Energietechnik is responsible for climatic design. One of the first steps for a climate engineer is to carry out an analysis of the environmental situation of the buildings and their orientation. The next stage is to examine the cultural background, taking a closer look at the standard types of urban design and buildings in the region. In terms of their historical development, Dubai and Muscat represent the reference cities for Masdar City. Built in a comparable climate, they feature specific patterns: the buildings are separated by very narrow streets (practically footpaths) and the squares are shaded, factors that minimize the exposure to solar energy.

A range of strategies could be identified for Masdar City from the way that local flora and fauna have adapted to the environment of the United Arab Emirates: protection from the sun and dust, collection of dew for water gain and reliance on a minimum amount of water. All of this research formed the basis for the urban master plan.

A sustainable approach

Urban densification is the most important measure in the approach to sustainability in Masdar City. It makes the greatest contribution to maintaining low energy requirements in this hot, humid climate. It follows that the total energy consumption must be met by renewable energy sources, and that all of the materials used have to be recyclable. As the options for utilizing solar, wind and geothermal energy sources are limited, energy demand has to be reduced through technical solutions but also by lifestyle changes related to mobility, expectations of comfort, the consumption of water, energy and materials, as well as waste generation.

Figure 13.1 Artist's impressions of a Masdar square

Source: Foster and Partners

Developing guidelines on energy and comfort

Generic models for streets and buildings were developed as the basis for the urban design of Masdar City. These models, which allow for evaluation and improvement, were analysed using flow simulations and daylight and shade measurements. A similar approach was applied to the initial calculations of the dynamic building load. This involved using specific construction methods, internal and external loads and building utilization in the assumptions. The loads were recalculated on the basis of the generic building typologies that were developed; from this, the necessary adjustments were deduced to ensure that the energy required would not exceed the amount that could be produced on site.

Fresh air without heat

Banning cars with ICEs from Masdar City was a decision necessitated by the carbon neutrality requirement for the project. This reduced the need for

Figure 13.2 Masdar master plan

Source: Foster and Partners

mechanical ventilation in the city, which depends on, among other things, the so-called heat island effect, a phenomenon caused by solar radiation and urban emissions. Using an isothermal flow model, it was possible to determine the optimum layout – length and breadth – of streets.

In order to examine the influence of the high solar heat ingress produced by the photovoltaic roofs on the microclimate above the city, a simplified model of a whole street was created. The results allowed conclusions to be drawn about the layout of the streets and squares. The use of detailed models made it possible to work out principles for producing a 'cold island effect' to lower the ambient temperature. Additionally, a reinterpretation of the

Emirates' traditional wind tower was created, which protects the street from solar radiation and wind and can be used at the same time as a form of ventilation.

Figure 13.3 The reinterpretation of an Emirates traditional wind tow and its possible impact were studied using fluid dynamic simulation

Source: Foster and Partners

The air quality requirements can be met and the local climatic conditions optimized by using the proposed layout and dimensions for the city grid. Two strips of parkland will run the whole length of Masdar City. These essential fresh air corridors will allow both sea breezes and the cool night wind to permeate the city.

Light without shade

Because of the high irradiation values in Masdar City, the interior and exterior of buildings have to be shaded perceptibly, while not being too heavily blacked out. Street widths, diameters of inner courtyards, building heights and the proportion of glass facades were all calculated with the aid of sun path diagrams, shade research and generic building models. All of the values were compiled in a matrix in order to facilitate the development of building typologies and their configuration. Finally, extreme conditions for direct solar irradiation were calculated and analysed. The interpretation of the results led the design team to conclude that daylight provision for the buildings should be ensured via the inner courtyards, and not through apertures in the street facades. This means that the streets can be narrow, which makes it easier to achieve a thermal effect. All the inner courtyards need either flexible forms of shading, in order to protect the microclimate from high temperatures, or externally situated sun screens on the openings to the courtyard facades.

Building loads in the city

The layout of the city affects its climate, making it an important boundary condition for the building load. This in turn is determined by the configuration of the buildings, the construction style and the user-dependent internal and external loads. This meant that each building typology required a separate form of modelling, which took account of the internal user profiles and detailed weather data. The values for air conditioning and electricity demand were identified using building simulation software and generic models. The boundary conditions were incorporated in three iterative steps and the final building typologies were specified. The resultant values for the overall loads were also important guidelines for the design of the central systems and utilities.

Planning changes

The original design of Masdar City consisted of two square areas: the larger one was to be the urban zone and the smaller strictly a service centre. The decision to use the smaller square for buildings as well meant that a large section of the surface area designed for power generation was lost. A fourth iterative step was then needed as a result of these restrictions to the energy supply. Consequently new boundary conditions had to be established for the building design and fittings, and additional guidelines for users defined. The energy requirements had to be reduced in line with the decrease in the amount of energy produced.

Limits: 80% reduction in demand

The target is an 80 per cent reduction in energy demand compared to the current level in the United Arab Emirates. This has serious consequences for the building typology – offices, laboratories, residential housing, retail outlets and light industry units etc. The following three steps lead to sustainable urban development based on carbon-neutral operations: reducing primary energy consumption by 40 per cent through passive design techniques; optimizing supply systems and energy consumption strategies, making a further saving of 30–40 per cent; and finally, lowering the remaining 20–30 per cent of primary energy consumption through renewable forms of energy and active recycling efforts.

Ambition

Once the design has been realized, the hope is that Masdar City can act as a flagship that will inspire future imitation and thus have widespread impact. The sponsors of the concept hope that such 'carbon-neutral cities' should be reproduced on a global scale.

Credits

Client: Masdar-Abu Dhabi Future Energy Company, Mubadala Development Company

Business plan: Ernst & Young

Urban design and architecture: Foster and Partners, London

Renewable energy: E.T.A., Florence

Cimate engineering: Transsolar Energietechnik GmbH, Stuttgart

Sustainability infrastructure: WSP Group plc, London

HVAC engineer: WSP

Transportation: Systematica, Milan

Quantity surveyor: Cyril Sweet Limited

Landscape consultant: Gustafson Porter

Site area: 600 hectares

Occupied land: 296 hectares

Floor space: 4.8 million m² plus optional 1.2 million m² (second square)

Chapter Fourteen

Urban Energy Potentials: A Step Towards the Use of 100% Renewable Energies

Dieter D. Genske, Lars Porsche
and Ariane Ruff

Today, attempts are made to develop sustainable urban areas that utilize only renewable energies and thus contribute efficiently to the reduction of greenhouse gases. The example of Masdar City in Abu Dhabi, presented in Chapter 13, is an example *par excellence*, ready to cope with all environmental challenges. However, Masdar City also still reads as a mirage – a single showcase of the possible – while countries must also urgently cope with the primary challenge: the introduction of renewable energies in existing cities. These cities have developed over many centuries and where founded at times when climate protection was not an issue. Paris, London and Berlin cannot simply be rebuilt. Instead, smart changes in existing cities have to be introduced, utilizing existing buildings and infrastructure and respecting historic building substance. This task appears to be the real challenge, especially since it has to be achieved in a participatory way, allowing citizens to take part in and to profit from it.

The production of renewable energy needs space, which is rarely available in cities. Is it? Our research has shown that, in fact, a lot of space is available, even in a densely populated urban environment, and that these spaces may well be utilized to produce renewable energy. They include derelict urban terrain, roofs and facades, watercourses and wastewater, ambient air and exhaust air as well as the underground. It is only a matter of finding these spaces in order to identify their potential to produce renewable energy. Once these potentials have been determined, the potential for reduction of greenhouse gases can be calculated as well.

Since communities are rarely aware of their complete space potentials, they can hardly assess the opportunities these spaces bear. Due to the variety of renewable energy production strategies, these spaces can, however, readily be utilized for energy production. By contrast, virgin land is consumed to produce energy from non-renewable resources, often far away from the city where the energy is needed. In addition, land is occupied for biofuel production that is urgently needed for food production. Energy production *extra-muros* leaves ecological and social footprints that will continue to grow and to exhaust the resources still available. Realizing the upcoming resource and climate crisis, many nations have launched national strategies of sustainability. In Germany, for instancc, land consumption is to be reduced to 30 hectares per day until 2020 while at the same time the share of renewable power generation should be increased to at least 20 per cent and greenhouse gases should be reduced by 40 per cent as compared to 1990. This means that used land, i.e. brownfields, but also so far unused blank spaces have to be screened for their suitability for renewable energy production.

This chapter focuses on four questions: Which options of renewable energy production are appropriate in an urban environment? How can they be combined to find smart synergies? How much green energy may eventually be produced in the city and its immediate surroundings? What will be the effect on the climate?

Renewable energy in cities

There is a large variety of options to produce renewable energies. They basically comprise the utilization of sun, water and wind as well as the subground and manmade sources such as exhaust air or wastewater. To this add the wide field of biomass utilization. Figure 14.1 gives a brief overview of the options of renewable energy production, distinguishing between options of power generation and options of heat supply.

Not all of these options suit an urban environment. The many criteria to judge their compatibility include energy production costs, spatial efficiency, availability, ecological impact, compatibility with the cityscape, acceptance by the citizen, robustness against vandalism, landmark protection, ability to be easily dismantled and legal complications. In the ExWoSt (Experimental Housing and Urban Development) study carried out by Genske et al (2009d) it became clear that the following options are suitable for the urban environment:

- Photovoltaics and solar collectors may be installed on roofs and facades.
- Waterpower can be generated from rivers, canals and pipes.
- Shallow geothermal heat can be exploited with ground source heat pumps.
- Ambient heat, i.e. exhaust heat and exterior heat, can easily be exploited to heat buildings
- Thermal energy can be extracted from wastewater already in the building, or from the public sewer system or from the cleaned wastewater after the sewage treatment plant.
- Biomass, especially biowaste, can be converted into power and heat. Co-generation of heat and power increases the efficiency of the system.

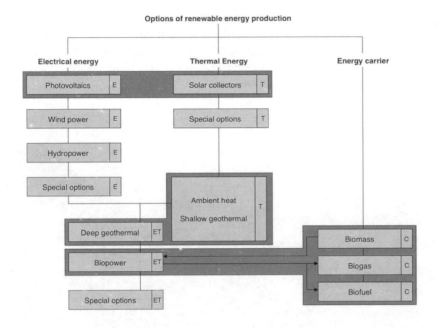

Figure 14.1 Options for energy production

Source: BBSR within the BBR
Note: E = electrical energy, T = energy, C = energy carrier.

The following options do not appear to be suitable:

- Large windmills are rarely erected in cities. They disturb the skyline and occupy too much space, since safety zones have to be respected. However, outside the city centre they may well be employed, as in historic times, before the exploitation of fossil resources made them obsolete.
- Small windmills are not cost efficient (so far), since in most cases the city profile is much too rough to allow for strong and stable winds. On roof edges turbulence makes it necessary to elevate the windmills, which again disturbs the skyline.

The following options depend on certain boundary conditions that have to be analysed from case to case:

- Free-range photovoltaic power plants may only be considered sustainable if the terrain cannot be used otherwise. Derelict spaces should be developed to make cities more dense and compact and thus reduce land consumption outside the city, instead of being blocked by photovoltaic panels. However, on landfills or highly contaminated sites, photovoltaic plants may be an option.
- Growing biomass in the city, especially energy plants and wood, only makes sense if combined with measures of rendering the city more attractive. This

may, for instance, be the case if a derelict terrain is regreened with biomass to cover an urban void that could deter potential investors (Dosch and Porsche, 2008). In Erfurt, Germany, energy plants are grown in the city centre as a means to encourage people to convert biomass into energy (see Figure 14.2) (Schumacher, 2008).

Figure 14.2 'Power plants' in Erfurt

Source: Ariane Ruff

A fundamental aspect is the investment associated with energy production. In addition, the efficiency with regard to space consumption appears to be crucial. In Figure 14.3 the costs of generating electricity are plotted across the urban space needed. The upper limits refer to current costs, the lower limits are cost predictions for 2020 (BMU, 2008). As can be seen, photovoltaics appear to be highly efficient but rather expensive, whereas biomass is cheap but fairly inefficient. A similar picture can be drawn for the options of heat generation (see Figure 14.4). Here, however, the costs for generating solar heat – solar collectors – are much lower. Rather inexpensive but still efficient are ground source heat pumps. It has to be kept in mind that these graphs refer to urban space. This means, for example, that photovoltaic panels are installed mainly on roofs and facades with urban restrictions as mentioned above already applied (Genske et al, 2009d). Because of this, optimal energy production achieved with free-range installations is only possible in certain cases.

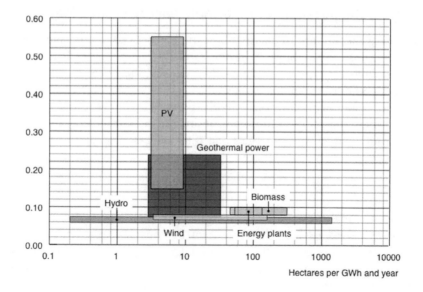

Figure 14.3 The demand on urban space to produce one GWh per year with options of renewable power generation and the associated costs

Source: BBSR within the BBR
Note: PV = photovoltaics. The upper cost limits reflect current prices, the lower ones are predicted for 2020

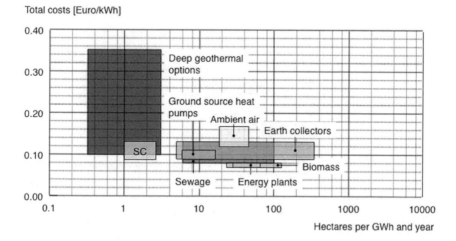

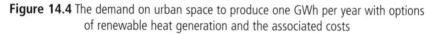

Figure 14.4 The demand on urban space to produce one GWh per year with options of renewable heat generation and the associated costs

Source: BBSR within the BBR
Note: SC = solar collectors. The upper cost limits reflect current prices, the lower ones are predicted for 2020

Assessing the potential of renewable energy production

Calculating the amount of energy that can be produced in an urban space is a demanding task. The ability to produce renewable energy depends on many factors such as the type of buildings, their density, the given infrastructure and the applicability of technologies as well as natural aspects such as annual solar radiation, geomorphology (for hydropower), wind speeds, geothermal potentials, etc. In addition, different options of renewable energy production can be combined to find synergies and thus increase and optimize the energy output.

Approach

In order to assess the energy potential, the city has to be divided into prototypic spaces such as 'historic centre', 'business districts', 'residential spaces' etc. (Everding, 2007). Table 14.1 lists the 14 urban prototypes that have been applied in this study. For these prototypes the energy demand has to be assessed. In this approach it is assumed that measures for energetic improvement have already been taken such as replacing inefficient windows, repairing the roof, revamping the facades, etc., i.e., common measures of renovation that are carried out once in a while. In Germany, renovations have to be carried out according to the Energieeinsparverordnung (EnEV), a legal framework introduced by the government to reduce the energy consumption of buildings (DENA, 2008). We are thus assessing the future energy demand, i.e., the typical demand for heating (and cooling) of buildings and the demand for electricity. Our projection refers to 2020.

Table 14.1 Urban prototypes

Class	No.	Subclass
Mixed types	I	Pre-industrial city/historic centre
	II	Building blocks of the 19th and early 20th centuries of the central city
	III	Post-war reconstruction (of destroyed buildings)
	IV	Village-like fractured structure
Living	V	Pre-war company housing
	VI	Social (subsidized) housing of the 1950s
	VII	High-rise apartment buildings of the 1970s and prefabricated block structures (mainly former East Germany)
	VIII	Apartment buildings since the 1960s
	IX	Private homes/residential areas (single-family homes)
Business	X	Business and industry
Functional buildings	XI	Service buildings, office buildings, shopping malls etc.
Park and green	XII	Parks, open air sports grounds, city forests, gardens, graveyards
Agriculture	XIII	Crops and animals
Rest	XIV	

Source: Based on Everding (2007)

Furthermore, four 'energy parties' are distinguished: households, businesses (trade, services, businesses), industry and mobility. Figure 14.5 indicates the relative energy demand of these four energy parties in Germany. Our study includes the first two energy parties (households and businesses). Finally, discrete and diffuse energy sources are distinguished. Diffuse energy sources neither cause material streams nor do they block any additional space, whereas concrete energy sources consume spaces and trigger material streams. For instance, photovoltaic panels do not occupy extra space if installed on roofs and facades and do not trigger any material streams while in operation. They are thus considered as a diffuse option. Growing biomass uses up spaces that cannot be utilized for any other purpose at that moment and is thus considered a concrete option. When analysing energy scenarios, only diffuse energy sources are considered in the first place that are eventually superimposed with concrete options that are already existing (for instance, a biomass power plant) or may be introduced to satisfy local demands.

In special maps, the capacity to produce renewable energy is compared with the energy demand. Since maps of renewable energy potentials are drawn for the whole city, spatial energy deficiencies can be identified and balanced with concrete options such as new power plants that rely on regional renewable energy resources such as biomass or geothermal power.

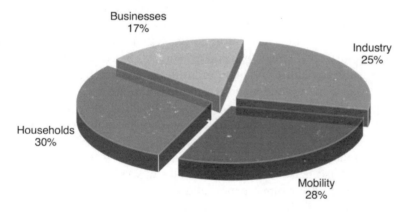

Figure 14.5 Structure of energy demand in Germany according to energy parties

Source: BMU (2008)

Results

This approach to assessing the potential of renewable energy production has been applied to a number of model regions including the cities of Stuttgart, Leipzig, Gelsenkirchen and smaller cities in Thuringia. The portfolio of model regions thus includes booming cities (Stuttgart), stable cities (Leipzig) and shrinking cities (Gelsenkirchen and northern Thuringia). Since all these cities are located in northern Europe, providing heat dominates the energy demand of the two energy parties considered (households and businesses). In fact, more than half of the energy consumed by a regular household is needed for heating and warm water supply, while only one fifth is needed for electricity and one third for mobility (BMU 2007). Heat cannot be transported over long distances whereas electricity is mainly independent of the point of input. This means that the heat demand has to be covered first in a decentralized way with renewable options. Only after the heat demand is satisfied can options to generate electrical power be considered. Practically, this means that at the moment a surplus of heat can be supplied, roofs and facades may be utilized to generate electricity with photovoltaic panels instead of heat with solar collectors, thus yielding more urban-generated electricity and increasing the degree of electric autarky.

The general strategy of deriving maps illustrating the potential of renewable energy generation is depicted in Figure 14.6. Based on data such as local statistics, historical maps, current city maps, aerial photos, service maps etc., the city is divided into urban prototypes (of which 14 are applied, see Table 14.1). For these prototypes, the long-term energy demand is assessed, taking into consideration that at least one cycle of renovation has already taken place. Thereafter, the corresponding potential to produce renewable energy is determined with the distinction made between diffuse and concrete potentials. Based on this, the ratio of renewable energy produced to energy needed is calculated for both thermal and electric energy. Finally, the outcome is visualized by means of a geo-information system. The resulting map is applied to optimize the renewable energy supply of the city by including existing energy infrastructure and planning additional concrete supply options.

In order to analyse a model region, typical scenarios are established. They reflect simple situations such as the sole introduction of solar collectors (scenario I) and the equally weighted introduction of solar collectors and photovoltaic panels (scenario II), more sophisticated scenarios that include further heat generating options (ambient and shallow geothermal heat, wastewater heat) (scenarios III–V) and finally a scenario that adjusts the prototypic share of photovoltaic panels and solar collectors to optimize heat generation (scenario VI). In Figure 14.7 the city of Sondershausen, Thuringia, is represented by means of the 14 urban prototypes and in Figure 14.8, scenario VI is shown, both for thermal and electrical potentials. Sondershausen has a population of 21,000 inhabitants and covers an area of 12,000 hectares. The city looses about 1.0 per cent of population every year and is thus qualified as a 'shrinking' city. High energy production can be observed in areas of large post-war apartment blocks, business quarters and modern residential areas with individual buildings (family homes), whereas the city centre with its historic buildings as well as urban building blocks of the 19th and early 20th centuries remain low in renewable energy production. In this scenario the thermal autarky is 98 per cent and the electric autarky is 93 per cent.

Scenario VII includes additional concrete options for energy supply. It takes into consideration a possible biomass co-generation power plant, the

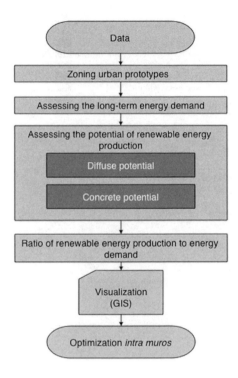

Figure 14.6 General strategy of deriving maps depicting the potential for renewable energy generation

Source: BBSR within the BBR
Note: GIS = geographical information system

extraction of heat at the outlet of the local sewage treatment plant and the extraction of heat from an old salt mining shaft. The degree of energy autarky can thus be pushed above 100 per cent, i.e. the energy demand of households and businesses can be satisfied with renewable resources *intra muros*.

In addition, the reduction of CO_2 emissions due to the introduction of renewable energy generation is assessed. For the city of Sondershausen, about 88,000t of CO_2 can be saved annually with scenario VII, i.e. some 2.5t per year per inhabitant, as a first estimate.

Future research goals

So far, only two energy parties (households and businesses) have been considered. The remaining two (industry, mobility) will be investigated in research projects that have just started (Genske et al, 2009a). Investigations into the energy party 'mobility' will include the idea of an electrification of the traffic. Battery powered vehicles will serve as energy buffer to store and trade regional surpluses of power production (V2G) (Engel, 2005). In addition, the energetic functions of the immediate surrounding of cities will be examined, thus turning the study from an urban into a regional one. In addition, the assessment of CO_2 reduction potential will be refined and also visualized with appropriate maps (Genske et al, 2009b).

The City of Sondershausen divided into urban prototypes

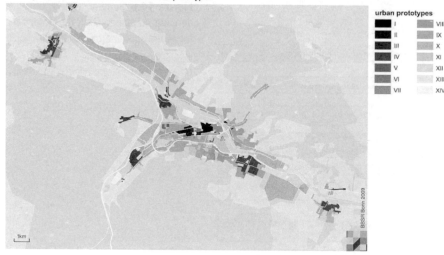

urban prototypes

■	I		VIII
■	II		IX
■	III		X
■	IV		XI
■	V		XII
■	VI		XIII
■	VII		XIV

1km

BBSR Bonn 2009

Source: BBSR, edit.: Nutzung städtischer
Freiflächen für erneuerbare Energien, 2009

Figure 14.7 The city of Sondershausen divided into urban prototypes

Source: BBSR within the BBR

Finally, a special research focus aims at identifying possibilities for financing the introduction of renewable energies. Special attention is given to the participation of citizens in this process. Formats of investment are investigated that let people participate in saving money and creating wealth with renewable energies.

Conclusion

Our studies have shown that it is possible to both assess the long-term energy demands for urban households and businesses as well as the potential of urban spaces to produce renewable energy. We have introduced a strategy to visualize these renewable energy potentials with maps covering the model region. Degrees of energy autarky have been calculated for thermal and electrical demands. The results are based on the analysis of six scenarios plus a seventh scenario that includes aspects of already existing energy infrastructure and new concrete measures of energy supply. For the model region of Sondershausen (Thuringia, Germany) it was found that enough renewable energy can be produced *intra muros* (within the city limits) to satisfy the long-term demands of households and businesses.

Although calculations for the other model regions are still ongoing, it can be stated that for bigger cities energy autarky cannot be achieved as easily as for small cities. Nevertheless, in most cases this appears to be still possible.

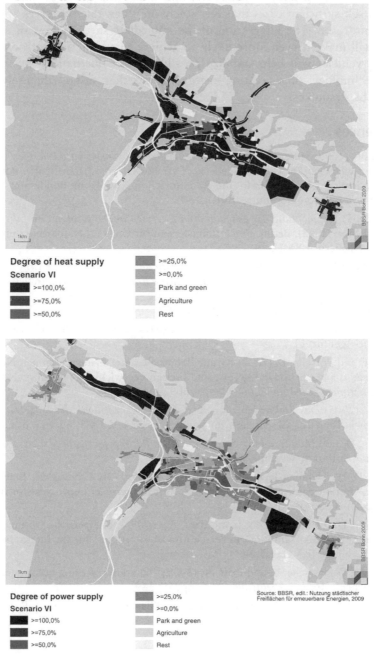

Degree of energy supply to satisfy long-term demands for the City of Sondershausen

Figure 14.8 Degree of energy supply to satisfy long-term demands for the city of Sondershausen

Source: BBSR within the BBR
Note: Depicted is the ratio of renewable heat generation to demand (top) and renewable power generation to demand (bottom). The darkest shades indicate a high potential to provide renewable energy

With increasing prices of fossil fuels and improving green technologies, the goal of self-sufficient cities that rely solely on renewable resources appears to be within reach. The integration of the immediate surroundings of an urban space will make it even more likely to achieve this goal. The introduction of electric vehicles will help stabilize possible power fluctuations by means of V2G technology, thus creating energy regions that sustain themselves with renewable resources.

References

BBSR/BBR (2009) *Nutzung städtischer Freiflächen für erneuerbare Energien* (edited by Dieter D. Genske, Thomas Joedecke, Lars Porsche and Ariane Ruff). Bonn, Bundesinstitut für Bau-, Stadt- und Raumforschung (BBSR) im Bundesamt für Bauwesen und Raumordnung (BBR), 140

BMU (2007) *Entwicklung der erneuerbaren Energien in Deutschland im Jahr 2007* (Stand März 2008), Berlin, Bundesministerium für Umwelt, Naturschutz und Reaktorsicherheit BMU, 24

BMU (2008) *Leitstudie 2008 Weiterentwicklung der 'Ausbaustrategie Erneuerbare Energien' vor dem Hintergrund der aktuellen Klimaschutzziele Deutschlands und Europas*, Berlin, Bundesministerium für Umwelt, Naturschutz und Reaktorsicherheit BMU (Bearb. J. Nitsch), 99

DENA (2008) *Auf einen Blick: Der Energieausweis kompakt*, Berlin, Deutsche Energie-Agentur, 11

Dosch, F. and Porsche, L. (2008) 'Grüne Potenziale unter blauem Himmel. Neue Zugänge zur Flächenrevitalisierung und Freiraumentwicklung im Ruhrgebiet', in *Informationen zur Raumentwicklung*, vol 9, pp609–625

Engel, T. (2005) *Das Elektrofahrzeug als Regelenergiekraftwerk des Solarzeitalters.* München, Deutsche Gesellschaft für Sonnenenergie e.V. DGS, 6 (www.dgs.de)

Everding, D. (ed) (2007) *Solarer Städtebau: Vom Pilotprojekt zum planerischen Leitbild*, Stuttgart, W. Kohlhammer

Genske, D. D., Henning-Jacob, J. and Ruff, A. (2009a) '3E: Erneuerbare Energien für Städte - Ein interaktives Expertensystem', Research project in progress for the European Union (EFRE-TNA)

Genske, D. D., Henning-Jacob, J. and Ruff, A. (2009b) 'Energetische Optimierung des Modellraums IBA Hamburg', Research project in progress for the Internationale Bauausstellung IBA Hamburg

Genske, D. D., Porsche, L. and Ruff, A. (2009c) 'Energieerzeugung auf urbanen Freiflächen', *Altlastenspektrum*, vol 6, pp259–269

Schumacher, H. (2008) 'Power plants': Energie-Pflanzen-Garten, Erfurt, FH Erfurt (Flyer)

Chapter Fifteen

Closing the Planning Gap: Moving to Renewable Communities

Nancy Carlisle and Brian Bush

Many communities fail to meet their conventional energy reduction/climate change goals (Bailey, 2007) because they focus on short-term, incremental approaches instead of tackling the more challenging task of guiding the deeper transition to a 'renewable energy community'. As a result, even in many of the most engaged communities, growth in energy use is increasing at a faster rate than their savings in energy. Closing this 'planning gap' soon through greater use of renewable energy is critically important to move to significantly more energy efficient communities that are on target to meet climate change goals, and imperative for those leading edge communities with the goal of meeting all their direct energy needs.[1]

Incremental energy efficiency improvements are a necessary step to move to 100 per cent, but should be only one step in a comprehensive overall approach. Long-term planning is needed for communities to meet their needs with increasing fractions of renewable energy. Implementation will require a combination of strategies that include the application of renewable energy technologies and energy efficiency practices; linking values in support of reducing carbon-based energy use to human behaviour; and public policy to further encourage greater use of renewable energy and reduction in greenhouse gas emissions (which also address scarcity of resources and social equity issues).

Figure 15.1 characterizes what we refer to as the planning gap. This is defined as the gap between what a community needs to do to significantly increase their use of renewable energy and the reality of what is currently being done to actually respond to this need. Many communities develop goals to promote greater use of renewable energy; yet their actions tend to focus on near-term incremental approaches to improve energy efficiency rather than a staged and fundamentally new approach to achieve far greater levels of energy

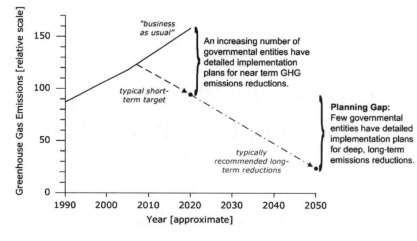

Figure 15.1 The planning gap

Note: GHG = greenhouse gas

efficiency and, particularly, integrated renewable energy. Community growth in energy use is increasing at a faster rate than its savings in energy (Brown et al, 2008).

In this chapter, we provide a working definition for a 100 per cent renewable community, discuss the reasons why the move to a large fraction of renewables is challenging for communities, provide examples of approaches that various communities are taking to move toward greater reliance on renewable energy, identify lessons learned from each example and describe the common elements found in the most successful approaches, as messages to communities who wish to make fundamental improvements to their practices. We conclude with steps needed in terms of research.

100% renewable community: A working definition

The word 'community' is used in this chapter to mean a group of people of any size that occupies or uses buildings in a specific location, has a shared focus and some form of governance. A community could be a neighbourhood, a subdivision, a planned community, a military base, a university campus, a research park, or a city or town. One challenging issue is defining the community boundary.

Our working definition for a 100 per cent renewable community is one that uses renewable energy generated on-site or from a resource owned by the community offsite or meets 100 per cent of the 'direct uses' of energy within the boundary of the community for all buildings, community infrastructure (energy for water, waste, lighting, etc.) and transportation systems. It is assumed that these needs are met first through the use of energy efficiency, and once energy efficiency is fully exploited, supplemented with renewable sources (first onsite and then offsite). The timeframe to reach 100 per cent can be phased as long as it is defined. We have excluded indirect or embodied forms of energy within the community from the definition only because of the challenge of measuring these at this point in time. Indirect uses of energy within a community include energy for growing food, transporting goods and energy embodied in materials and services.

The energy used for the community is measured by comparing a baseline of energy needs for buildings, community infrastructure and transportation for the community to the energy supplied on an annual basis from renewable energy sources. Measuring is done on the basis of BTUs (British thermal units) in order to compare all renewable energy supplied (for electricity, thermal and transport fuels) to all loads to determine if the goal was achieved over a given time period. In this way, for example, excess renewable electricity production could be used to offset the use of conventional thermal energy or transportation fuels used.

We can use the hierarchy for a grid-connected zero-energy building (ZEB) as defined by Torcellini et al (2006) as a starting point for valuing renewable energy options, not only for buildings but also for community-scale applications. The hierarchy defines the relative value of various sources of renewable power to serve a community. A ZEB is defined as a residential or commercial building with greatly reduced energy needs through efficiency gains such that the balance of energy needs can be supplied with renewable technologies.

It is more challenging to define metrics for a zero-energy community than a building because energy in communities is used for community infrastructure and vehicles as well as buildings, and there are more types of fuel to consider. It is a challenge to define what community energy uses are included in the net zero calculations, the boundary for the community and the timeframe necessary to meet the goal.

Table 15.1 follows the logic established by Torcellini et al (2006) in applying the hierarchy for a net ZEB to the definition of a net zero-energy community. The table includes options for transportation, community infrastructure and modifying human behaviour. Option number zero (0) includes strategies to reduce the energy load through either energy efficiency and\or energy conservation strategies. These are the lowest cost strategies. Supply-side strategies, beginning with Option 1, represent ways to use renewable energy as a source of energy. Table 15.1 could be further quantified by assigning multipliers to each source of renewable supply as a way to assign value. For example, a solar thermal or electrical system located on a building rooftop has a higher multiplier than the value of renewable electrical system located near, but not on, the buildings due to the higher line losses from a renewable system that is near the building. Likewise a biomass boiler serving a community would have a lower multiplier to account for the transport of the biomass fuel to the site.

Table 15.1 Conventional energy reduction options hierarchy for a community

Option number	Demand- and supply-side options	Examples
0	Reduce building and infrastructure site energy use through low-energy building technologies for new construction; energy retrofits for existing buildings and reduce vehicle miles travelled for vehicles.	*Buildings technologies* – use of daylight, high-efficiency HVAC equipment, natural ventilation, evaporative cooling, insulation etc. Aggressive energy efficiency standards for all construction. *Behaviour* – Turn lights and equipment off at night, metering, commissioning and retro-commissioning; provide customers with feedback on power usage; turn off vampire loads in buildings at night and when buildings are not occupied. *Vehicles* – Use alternative transportation, advanced vehicles and fuels; maintain and operate vehicles to maximize vehicle efficiency; form car-sharing clubs; provide and plan for alternative transportation and access to mass transit within the community. *Infrastructure* – LED traffic lights, high efficiency pumps. Utilities are installing smart grids to provide user feedback regarding energy use and impact on their energy costs. *Caveats* – Plug loads in homes and offices are constantly growing; smart grid monitoring always on and, therefore, increase overall loads.
Onsite Supply Options		
1	Use renewable energy sources available within the building's footprint as the first priority.	*Buildings* – Photovoltaics, solar hot water, ground source heat pumps located on the building and connected to building systems. *Behaviour* – Consumers with renewable systems located on their buildings tend to conserve more because they are more aware of how they use energy. *Vehicles* – Use electric or (when available) plug-in hybrid vehicles from on-site-generated renewable energy and/or use alternative fuels in a flex fuel vehicle or use electricity from grid to charge the vehicles when renewable energy is unavailable. *Infrastructure* – Use renewable energy to power street lights, monitors and meters. *Caveat* – In order to maximize roof area for solar, orientation and street layouts, need to be considered; passive solar design features can significantly reduce heating, air-conditioning and lighting loads.
2	Use renewable sources available within the community boundary and connected to the building electrical or hot water distribution.	*Buildings* – Photovoltaics, solar, low-impact hydro or wind located on open space, parking lots etc. and the renewable energy output connected to the buildings electrical or hot water system. *Infrastructure* – Community scale micro-grid connects distributed and community-scale renewable systems (electric and or thermal energy) to buildings and utility grid. Infrastructure needs to address energy storage onsite or though grid-back-up. *Vehicles* – Cars will soon be sold with the ability to charge vehicles onsite, which can take advantage of using renewable energy generated onsite.

Table 15.1 continued

Option number	Demand- and supply-side options	Examples
		Vehicles will someday be able to provide power to the house if necessary in a back-up utility mode.
		Power system for home and vehicle charging will someday be interconnected.
Offsite Supply Options		
3	Use renewable energy sources available offsite for energy use onsite.	*Buildings and Community infrastructure* – Biofuels or landfill gas that can be imported from outside the community, or waste streams from onsite processes that can be used in the community to generate electricity and heat. A community could also negotiate with its power provider to install dedicated wind turbines or photovoltaic panels at a site with good solar/wind resources outside the community. In this approach, the community might own the hardware and receive credits for the power. The power company or a contractor would maintain the hardware.
		Waste-water treatment can be designed to minimize energy use and create energy from anaerobic digestion to capture methane for community heat and power.
		Vehicles – Use alternative fuels whenever possible.
4	Purchase NEW offsite renewable energy sources – RECs that result in additional generation added to the grid.	RECs typically include utility-based wind, photovoltaics, emissions credits or other 'green' purchasing options. Hydroelectric is sometimes considered. RECs could be used as a strategy to meet a goal for an interim period of time or as a 'top off' strategy to provide, for example, the last 10 per cent of renewable energy. It is important that REC purchase add new generation capacity to the grid.

Accounting for embodied energy

Our working definition of a 100 per cent renewable energy community currently just includes accounting for direct energy uses. However, in the future, we can further reduce community energy use by accounting for both direct and indirect energy use – energy embodied in the community's imported goods, such as food and water. A community might also have the goal of zero waste, possibly using a combustion process to dispose of waste, in addition to recycling and reducing; zero net water use is another related goal. All of the indirect uses of energy are challenging to account for and will require more research and agreement on consistent methodology and accounting principles. However, Lenzen et al (2008) propose a direct accounting method, based on household expenditures, that appears to form a sound base for planning. One approach might be to take a regional energy strategy versus a community strategy so that communities can be interdependent in terms of trade.

The planning gap: Goal setting to reduce energy and emissions on the rise

As the awareness in the US increases about carbon-based fuels issues and global climate change, cities, states, universities and federal agencies adopt far-reaching goals to reduce energy use and greenhouse gas emissions. For example, as of 20 May 2008, more than 547 university presidents had signed on to the American College and University Presidents Climate Commitment (2008) (www.presidentsclimatecommitment.org/html/commitment.php). It states:

> We recognize the scientific consensus that global warming is real and is largely being caused by humans. We further recognize the need to reduce the global emission of greenhouse gases by 80% by mid-century at the latest, in order to avert the worst impacts of global warming and to reestablish the more stable climatic conditions that have made human progress over the last 10,000 years possible.

Similarly, as of 20 May 2008, 852 cities had signed the US Conference of Mayors (2005) Climate Protection Agreement (www.usmayors.org/climate protection/agreement.htm) calling on mayors to meet or exceed the Kyoto Protocol targets to reduce global warming pollution levels to the reduction target suggested for the US of 7 per cent below 1990 levels by 2012.

Hundreds of architecture firms and other individuals have adopted the Architecture 2030 (2008) Challenge (www.architecture2030.org/2030_challenge/index.html), which calls for the global architecture and building community to adopt the following targets:

- All new buildings, developments and major renovations shall be designed to meet a fossil fuel, greenhouse gas-emitting, energy-consumption performance standard of 50 per cent of the regional (or country) average for that building type.
- At a minimum, an equal amount of existing building area shall be renovated annually to meet a fossil fuel, greenhouse gas-emitting, energy-consumption performance standard of 50 per cent of the regional (or country) average for that building type.
- The fossil fuel reduction standard for all new buildings shall be increased to 60 per cent in 2010, 70 per cent in 2015, 80 per cent in 2020, 90 per cent in 2025 and carbon neutral in 2030 (using no fossil fuel greenhouse gas-emitting energy to operate). These targets may be accomplished by implementing innovative sustainable design strategies, generating onsite renewable power, and/or purchasing (20 per cent maximum) renewable energy and/or certified renewable energy credits(Architecture 2030, www.architecture2030.org/2030_challenge/index.html).

These goals are an important step to begin the process as communities move to meet a significantly larger proportion of their power from renewable sources with 100 per cent of the direct power needs met by renewable energy. This is the ultimate accomplishment for communities on the leading edge. The challenge is that energy use in most communities continues to rise, not decline due to growth, new construction and/or greater needs for electric loads such as home electronics and air conditioning. Reaching a large fraction of a community's power from renewable energy will require a solution that combines a focus on technology and policy improvements with a fundamental shift in human behaviour and lifestyle.

The planning gap: Goals not being reached

Currently there is a gap between the goals to use significantly more renewable energy in communities and the reality. For example, the Colorado Climate Action Plan of November 2007 (Ritter, 2007) envisions radical reductions in state-wide greenhouse gases by 2050. The plan is built on the assumption that it is important to build bridging strategies to reduce greenhouse gases today while developing cleaner and less costly renewable technologies for the future. The plan focuses most of its emphasis on meeting the near-term target of 20 per cent emissions reduction by 2020 but devotes less than one page to getting from the 2020 to the 2050 greenhouse gas reduction target of 80 per cent. The example illustrates the general situation that communities tend to focus on reducing conventional energy use in the short-term instead of planning now for the more challenging task to guide the deeper transition to a zero-carbon energy economy. In contrast to this typical short-term planning, the approach for the farther time horizon with more radical goals will require a more strategic perspective, addressing fundamental technological uncertainties, complex financing mechanisms, and the need for behavioural change. This is especially true regarding energy infrastructure, which has long-lasting impacts or decisions regarding new construction where the decision to build in the most highly energy efficient manner makes the most economic sense when viewed from a lifecycle perspective; but is more typically viewed from a short-term or first-cost perspective.

100% renewable communities require an understanding of decision-making trade-offs including human behaviour

An implementation strategy to embrace 100 per cent renewables for communities needs to embrace both technological solutions as well as changes in behaviour, and do so in a financially viable manner. Behavioural understanding and decision-making are key elements that are not always considered in debates about strategies to move to 100 per cent renewable technologies and in discussing climate change policy. While beyond the scope of this chapter, studies in the field of behavioural economics by Little (2006), Lambert (2006), Sherman and Booth Sweeney (2006) and Moezzi (1998) look at adult belief systems, the connection between beliefs, public policy and behaviour. While there are many papers that define technical solutions to solve our climate-impacts-mitigation problems using efficiency and renewable energy, there is little help for community implementation: looking at financial structures and human behaviour to ultimately lead to adopting these new technologies.

One tool does support this broader approach. At the National Renewable Energy Laboratory in Golden, Colorado, Dr A. Walker (2008) has developed a method to analyse renewable energy options for a specific site, which can provide the basis for stakeholder discussions to determine necessary buy-in from the community level, and a finance strategy and implementation plan.

Using his renewable optimization software Walker first identifies the least-cost combination of renewable energy technologies to provide 100 per cent net renewable power for a specific site. At the National Zoological Park in Washington DC and the 4600 acre Conservation Research Center in Fort Royal, VA, he identifies the lifecycle cost of the combination of renewable energy technologies that would meet their goal of 100 per cent net renewable energy by the year 2016 at $74 million versus the lifecycle cost of continuing to purchase gas and electricity, which has a lifecycle cost of $52 million. The lifecycle costs do not include a dollar value for pollution emissions, educational value or other benefits associated with using renewable energy (Walker 2008). This analysis provides a launching pad for stakeholder discussion for the community to determine how best to meet this goal. Identifying the lifecycle cost of both options provides a basis to develop a financing plan. Without understanding the magnitude of the investment required and without considering the other benefits offered by the net zero solution, one might not realize the importance of developing a long-term financing plan. This tool has been used for specific business and government campuses and small towns including the Smithsonian National Zoo, Frito Lay, Anheuser Busch, Greensburg, Kansas and San Nicholas Island – an island owned by the US Navy.

Closing the planning gap: Community planning examples

We have grouped community planning examples into three categories. The categories describe various approaches to reducing conventional energy use and embracing renewable energy. The categories range from taking incremental steps to more visionary and aggressive approaches. Within each category we cite example communities and lessons learned from the approach taken.

Target incremental savings

Many local communities and university campuses (essentially small communities) have developed action plans to reduce their greenhouse gas emissions and move towards the concept of greater reliance on renewable energy but their plan strategies and tactics focus on near-term incremental actions designed to improve energy efficiency only with little focus on the longer-term strategies. In many cases growth in their energy use rises faster than their savings. Incremental improvements that succeed in reducing emissions over the next decade cannot be repeated indefinitely to achieve ever lower emissions levels. Furthermore, once the 'low hanging fruit' of existing practices has been 'picked', additional progress will require longer planning horizons and more complex analysis and decision-making. Incremental energy efficiency improvements are a very necessary step to move to 100 per cent, but should not be the sole vision; rather they should be one step in a comprehensive overall approach to reach 100 per cent renewable energy. Incremental improvements made in the absence of a long-term strategy pose the risk of making particular further improvements more costly or otherwise impractical.

Example: The City of Boulder, Colorado
Boulder, Colorado, is a city of 100,000 and home of the University of Colorado. The Boulder Climate Action Plan focuses on the near-term goal of incremental improvements to meet a 2012 goal yet is silent on defining a strategy for a longer-term renewable goal (City of Boulder, 2006). In Boulder, as in most US cities, carbon emissions have grown since 1990. Brown et al (2008) state that carbon emissions in the 100 largest metropolitan areas in the US have increased by 1 per cent per year since 1908. Given the growth in energy use in Boulder, in order to meet the Kyoto Protocol reduction target of 7 per cent below the 1990 level, they now need to reduce emissions by approximately 24 per cent between 2005 and 2012. Figure 15.2 shows their total increase in greenhouse gas emissions between 1990 and 2005, with projections to 2012.

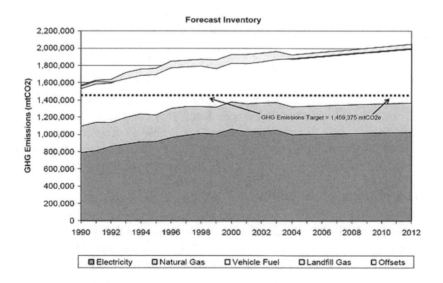

Figure 15.2 Growth in energy use in the City of Boulder

Source: City of Boulder (2006)

Between 1995 and 2007, the size of a new home in Boulder County has increased from about 2700 square foot to more than 6000 square foot (Kelley, 2008). These homes are more energy efficient than homes built in the past, but typically use a broader range of equipment such as air conditioning and electronic equipment. Despite impressive efficiency gains, the total energy used in buildings in the US almost doubled between 1970 and 2005; and the nation can expect to see building energy consumption increase by 0.8 per cent per year through to 2030 (Brown et al, 2008). This is coupled with the added dilemma in Boulder that housing prices near downtown continue to rise, forcing people seeking affordable homes to move farther from their jobs, which in turn increases vehicle miles travelled to and from work unless a mass transit opportunity is in place. This cycle causes climate-impacts mitigation to worsen.

The Boulder Climate Action Plan (City of Boulder, 2006) began with an inventory detailing the sources of greenhouse gases. By establishing a city-wide carbon tax in 2006, voters demonstrated their desire to take local action to mitigate the impacts of climate change. The carbon tax for residences is currently $13 per household per year with a planned increase to $19/household/year. The tax funds the city's effort to reduce greenhouse gas emissions at approximately $1 million per year. The tax does not represent the environmental costs associated with carbon impacts.[2]

The energy efficiency and renewable recommendations in the Boulder plan are largely voluntary and the city's role is largely as a facilitator, educator and an advocate. The plan assumes that over 95 per cent of the costs to improve efficiency in the commercial, industrial and residential sectors will be borne by the private sector and the local utility company and that utility rebates will cover between 25 per cent and 50 per cent of the implementation cost of efficiency measures. It makes the assumption that very little city investment will be required. The renewable energy strategy is to buy renewable energy credits rather than building onsite renewable energy systems.

The Boulder Daily Camera (Snider, 2008) noted that Boulder will only make it about half-way to its goal of cutting enough greenhouse gases to comply with their Kyoto Potocol target. Boulder is cited as a case study in this chapter because the reasons that Boulder is not able to meet the Kyoto target are representative of what one may find in many other US cities that are both proactive and committed to meeting greenhouse gas-reduction targets.

Smart grid initiative in Boulder

In 2008, Boulder was successful in receiving an award from its power provider, Xcel Energy, to plan for a smart grid. With leveraged funding from various government grants, this could be up to a $100 million effort. The funding for this initiative dwarfs the funding of the Boulder Climate Action Plan and could result in improved energy efficiency. The term smart-power grid refers to a power grid that enables real-time communication between the consumer and the utility allowing the utility to optimize a consumer's energy usage based on environmental and/or price preferences (Xcel Energy, 2008). Through this award, the utility intends to install equipment upstream of consumers, in substations, to boost grid intelligence and reliability, with the intent of squeezing out some of the inefficiencies that push up costs. Consumers will see installation of advanced meters in their homes. The meters will be capable of two-way communication and will provide a gateway to allow the homeowner – or, with permission, the utility – to remotely control furnaces, lights, air conditioners and other devices.

The meters could also give the utility the ability to dynamically price electricity. In the future, this could allow Xcel to assign different prices to electricity used at different times of the day, or for electricity used for different purposes. It could charge one price for basic lighting and another for running air conditioners or recharging batteries of plug-in hybrid electric vehicles, which are expected to be marketed in three years or so. These pricing plans would require approval of state utility regulators, something Xcel has not yet sought.

The smart grid could be a first step to position the power company to shift its role from that of a producer of a commodity (electricity) to one of a distributor and manager of the flow of electricity based on real-time pricing (Reuyl, 2006). As Reuyl explains, from the point of view of a sustainable community that is more reliant on energy efficiency and renewable energy, this fundamental shift is very significant. In a community where the utilities have real-time information regarding dispatchable energy and a source of multiple small-scale distributed sources of generation such as photovoltaic rooftops, the utility role could evolve to managing, tracking and matching energy supply versus demand to insure a reliable and efficient utility system. With real-time pricing, home owners and businesses can be motivated to install photovoltaics onto their rooftops.

Lessons from this case study include:

- The overall community greenhouse gas and energy use baseline is an important first step to define and understand the sources of energy use.
- The plan does not define a vision (including a financing plan) for significantly increasing the use of renewable energy within the community.
- The voluntary plan does not seem to be successful at changing human behaviour to the degree necessary to overcome the impacts of growth and lifestyle that continue to drive up total energy use.
- The smart grid project has the ability to significantly change human behaviour in response to both feedback on energy use and price signals.
- The smart grid project might have a larger impact on reducing greenhouse gases than the Climate Action Plan and therefore should be integrated within the latter (City of Boulder, 2006).

The key points are similar to the findings of a survey conducted in ten cities by the Institute for Local Self-reliance regarding how well the cities were meeting their climate change goals and what strategies and methodologies they were using. They concluded that many cities will likely fail in their attempts unless complementary state and federal policies are put into place (Bailey, 2007).

Some of the lessons to be learned identified by Bailey (2007) include:

- The methodologies and assumptions to create greenhouse gas inventories among communities are not standardized – making comparisons between communities problematic.
- Community-wide emissions have risen since 1990, sometimes dramatically and it is unlikely that more than one or two of the ten cities studied and quite possibly none will reduce their greenhouse gas emissions to the level of the Kyoto Protocol (7 per cent below 1990 levels by 2012).
- Almost all the cities surveyed were expecting to realize a significant portion of their greenhouse gas reductions as a result of actions taken by higher levels of government. Relying too heavily on strategies out of the city's direct control could stunt local solutions and inhibit the city's investment in energy-related projects that have ancillary economic and environmental benefits.
- Cities are not investing significant amounts of their own money to reduce greenhouse gas emissions.

Visionary approaches: Phased and pragmatic

The examples cited below are termed visionary because in all cases they focus not only on incremental approaches to energy efficiency but include both some type of 'big idea' and an implementation strategy that can be phased to get to the desired end point (in one case a ZEB, in another a zero-energy district (ZED)). In all cases their strategy attempts to go beyond the incremental improvements.

Example: Fort Collins, Colorado

Fort Collins, Colorado is a town with approximately 130,000 people in northern Colorado. It is home to Colorado State University. The town and the university have started a programme called Fort ZED in order to 'transform the downtown and university district into a net ZED', meaning that the district would create as much thermal and electrical energy locally as it uses within its built environment. 'Local' is defined as within a 50-mile radius of the district. Not only does the power generated in Fort ZED need to be local, it also needs to be clean energy balanced with efficiency and conservation. Fort ZED is a strategic partnership that involves the city of Fort Collins, the municipal utility, the clean energy cluster (local businesses interested in technical leadership and creating economic development), and a stakeholder group that encompasses interest in linking the downtown, the university campus and the downtown river corridor.

Like Boulder, Fort Collins is a growing community. Between 1992 and 2006 the population grew by 54 per cent, energy grew by 70 per cent and peak demand grew by 105 per cent. In 2003, as a precursor to Fort ZED, Fort Collins developed an energy policy that set energy demand reduction targets and supply targets coupled with a renewable energy supply target:

- Reduce per capita energy consumption by 10 per cent by 2012.
- Reduce per capita peak demand by 15 per cent by 2012.
- Increase renewable energy to 15 per cent by 2017.

Figure 15.3 shows how the city of Fort Collins has quantified the need for renewable energy in order to meet a 10 per cent renewable energy target.

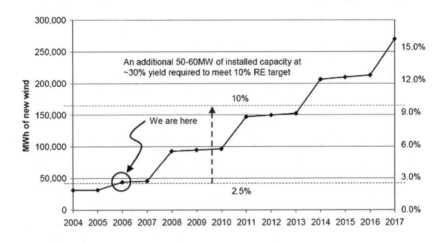

Figure 15.3 Fort Collins projections for renewable energy to meet defined energy targets

Source: Dorsey (2008)
Note: RE = renewable energy

They have determined that 50–60MW of installed capacity of wind would be needed to meet a 10 per cent renewable energy target. The programme has established strategies in five areas including:

- A 5MW (approximately 10 per cent of the district) demonstration programme. This involves identifying participants (such as businesses and university entities) among a cross section of the district to implement changes to achieve net zero goals. For example, the new Belgium Brewing company is a leader in the business community in terms of implementing strategies in support of zero energy goals. The Colorado State University is exploring the construction of a wind farm on land it owns in Fort Collins (an ambitious idea).
- The stakeholders in the district will develop a set of sustainability guidelines so that new construction projects and building retrofit projects designed will help to stabilize energy consumption in the district.

- The community will remove the barriers to develop needed infrastructure and create unique financial incentives in the district. This involves designing financial levers, tax incentives, incremental financing and design assistance as well as looking at the utility rate structure as a way to enable the desired outcome. The desired goal of this strategy is to affect local policy, regulation, markets and technology to facilitate Fort ZED.
- Fort Collins will work with their municipal utility to develop a strategy to use onsite renewable and distributed generation along with smart grid demand management to provide power to customers in the most efficient manner.
- Residents and businesses will be educated about this initiative and grassroots and neighbourhood groups will be encouraged to participate.

Lessons learned from this example include:

- Building strong stakeholder ties within the community in order to define the overall set of values that is desired in the community is an important first step. The shared set of values will help implement the portion of the strategy that impacts behaviour.
- The plan began with the vision and the high-level strategies for reaching the vision, along with some crucial early funding, such as being a recipient of a US Department of Energy grant to investigate peak load reduction using distributed and renewable resources. Fort Collins is currently doing a detailed analysis.
- The Fort ZED plan addresses the scale of the problem both in terms of technology and financial requirements and offers a time-phased approach to implementation.
- The plan involves the community members and uses them to 'model' behaviour as 'early adopters' for others to follow as an approach to address behaviour.
- The plan uses policy and incentives to shape the desired outcome.

Example: National Renewable Energy Laboratory building, Golden, Colorado

In a very general sense, a building is a community, although less complex. The National Renewable Energy Laboratory (NREL) is in the early design stages for a large (approximately 210,000 square foot) office building that has an energy design goal of 25,000BTU/ft²/year. The design of this building involves not only technology solutions but an understanding of the need to address human behaviour to reach its energy goal. Therefore, it is instructive as a case study because it is a pragmatic yet visionary approach to community scale, zero-energy solutions.

The first step in designing this building was to determine a measurable goal for energy efficiency: 25,000BTU/ft²/year of energy is approximately 50 per cent less energy than the minimum set under the ASHRAE 90.1 2004 standard

(ASHRAE, 2004). The intent is to use the goal as a tool to develop a comprehensive programme of energy efficiency measures and building operational strategies and policies to reduce energy use in the building as the first priority, rather than encouraging the use of supply-side renewable options coupled with a less efficient building where all energy efficiency options have not been fully exploited. This is an important strategy for communities as well as buildings. Renewable energy generation including photovoltaics, solar water heating, biomass, wind or renewable energy credits can then be used to take the energy use in the building from 25,000BTU/ft2/year to net-zero energy or 100 per cent net renewable energy.

Another step for this building is working with the occupants to turn off lights and equipment at night. A series of options were simulated for daytime versus night-time use of lights and equipment plug loads. For each pair of options, the total contribution to annual energy use was calculated. These analyses showed that if we operate the building at $0.8W/ft^2$ for plug load both during the day and night, the total contribution to plug load would be $18,000BTU/ft^2/year$ (Torcellini and Pless, personal communication, 2008). This analysis represents a case where no-one turns any lights or computers off at night. Obviously this would make it challenging to meet a $25,000BTU/ft^2/year$ building energy load if 72 per cent of the goal was being taken up by plug loads. As an alternative, if the building is operated at $0.8W/ft^2/year$ during the day and $0.2W/ft^2$ at night, the plug load contribution to the annual load can be reduced to $7000BTU/ft^2/year$. This is just one example of the importance of addressing human behaviour to achieve an energy saving goal. Addressing occupant behaviour is a shared responsibility of the design team and users. The design team can design in switches and easy ways to turn off the variety of loads associated with the buildings including such things as elevators and other building equipment.

Another option studied is building orientation and footprint. For the building to meet its efficiency goal, orientation and footprint matter. As seen in Figure 15.4, the building is being designed with a very narrow footprint, approximately 60 feet wide. In a commercial building approximately 40 per cent of the energy load is lighting, therefore, a strategy to reduce energy in commercial buildings is to significantly reduce the energy needed for lighting. Day lighting can best be accomplished from the north and south facades because these facades are easier to shade than those on the east and west sides. The long facades of the building are facing north and south in order to take advantage of natural lighting.

The intent is for the balance of the building load, the $25,000BTU/ft^2/year$ to be powered by a photovoltaic system, financed through a power purchase agreement, located onsite in order to bring the building energy to near zero. At the time of writing, the details of how much of the photovoltaic system can be located on the building versus the site has not been determined.

The key lessons that can be applied to communities include the following:

- As a first step, understand the end use (or drivers of) energy loads for a community.

Figure 15.4 Conceptual design image of the building

Source: RNL Design, Denver, CO

- It is important to set a measurable goal for the project or community and then break it into pieces to know what percentage can be met by efficiency, what percentage by onsite renewables and what percentage by offsite renewables.
- Exploit all the efficiency strategies as a first step and recognize that human behaviour and instilling an energy conservation ethic are part of the solution.
- Develop a plan and financing approach to meet a fraction of the goal using renewable energy.

Visionary and paradigm shifting

In this category, cities start with a bold vision and a financing plan. Developing a financing plan is an important step to take early on in the process. The examples are the county of Sonoma, California and Masdar City, in the heart of Abu Dhabi, United Arab Emirates. Masdar is under construction with a stated goal of being a 100 per cent renewable community.

Example: Sonoma County, California

The county of Sonoma, California recognized that financing is key to moving their community to a 100 per cent renewable energy community. At an early step in their process, they developed a financing strategy, now aggressively pursuing it. The county seeks federal funding to demonstrate its concept to provide enough local renewable power to meet the energy needs of a local business park, neighbourhood or segment of a town (Poole, 2008.) In justifying their argument for federal funding, the county cites many precedents that illustrate how the US federal government, involved in local partnerships, has acted as a catalyst for change in addressing issues of national importance.

The concept for which Sonoma is seeking funding is described here because it illustrates an example of a visionary idea that other cities might consider. Their vision includes four elements:

1   Developing a recycled water distribution system – their vision is to use a recycled water distribution system, geo-exchange wells and geothermal

heat pump technologies to heat and cool commercial buildings in urban and rural settings (i.e. winery operations and business parks, covering approximately 3 square miles or 3 per cent of Sonoma County's urban area). They would also make the regional distribution of recycled water available for irrigation and other non-potable uses. They estimate that this could save more than 50 per cent of the energy cost of water supplied for non-potable uses.

2   Using significantly more renewable energy – they also propose a programme to expand renewable power use by installing solar, wind, hydro-kinetic and landfill-powered generation systems in Sonoma County. These systems would provide sufficient local renewable power to meet the energy demands of a local business park, neighbourhood or segment of a city or town.

3   Developing a strategy for fuel efficient vehicles – they propose a demon-stration programme to address fuel efficient transportation. Their plan would be to seek a major auto manufacturer planning to produce PHEVs on a large scale and pursue incentive programmes with Pacific Gas & Electric and air quality districts to bring 100 to 250 affordable PHEVs to the citizens of Sonoma County.

4   Adding storage to their energy infrastructure – as a final element, their approach requires cost-effective, efficient systems for energy storage and recovery in order to supply power during periods when generation is low and demand is at peak. They will explore various technical options for storage. Their demonstration project would provide a means to test their concepts in 'real-life' applications in businesses, municipal agencies and communities.

This demonstration programme is championed by the Sonoma County Water Agency as a means to demonstrate the feasible implementation of a method and technology to retrofit existing communities that could be widely replicated to meet the power needs of the US, thereby reducing dependence on foreign oil and natural gas, as well as reducing water use and implementing 100 per cent reuse of recycled water.

The specific cost of implementing this concept in Sonoma County was not defined but they were requesting support for a $200 million block grant funding for local governmental projects focused on climate change (Poole, 2008).

Lessons from this case study include:

• Start with a big idea, plan for the long term and focus on a financing strategy.
• The use of some means to interconnect distributed sources of energy within the community and the need for storage is an implementation strategy on which very few communities have focused.
• Recognize and take advantage of the changing role of the utility.
• Their approach integrates several systems that typically are not interconnected – water, energy and transportation.

Example: Masdar City, Abu Dhabi

Masdar City is a new planned community being designed as a dense walled city that will cover 7km². It will be home to 50,000 people and 1500 businesses – primarily commercial and manufacturing facilities specializing in environmentally-friendly products. Another 40,000 workers are expected to commute to the city daily (Whittier, 2008) . The city's design uses traditional planning principals, together with energy efficient and renewable technologies available in the marketplace, to achieve a zero-carbon and zero-waste community. The concept for the city was initiated in 2006. It is estimated to cost $22 billion and will take approximately eight years to build, with the first phase scheduled to be complete and habitable in 2009 (Whittier, 2008). The city will be the location of a university, the Masdar Institute of Science and Technology, which is partnering with the Massachusetts Institute of Technology. Automobiles will be banned within the city. Travel will be accomplished via public mass transit and personal rapid transit systems, with existing road and railway connecting to other locations outside the city. No-one has designed a similar personal rapid transport system before. Masdar is being designed by the British architectural firm Foster and Partners.

Masdar will employ a variety of renewable power sources. Fifty-two per cent will be from photovoltaics (over 200MW), 28 per cent from concentrated solar (approximately 10MW); 14 per cent from evacuated tube; 7 per cent waste to energy, and exploration of geothermal is ongoing (Whittier, 2008). A renewable district cooling system, ~67,000t, coupled with a variety of desiccants (both liquid and solid) will be deployed on a centralized basis. The liquid desiccant system is particularly innovative for its low carbon footprint. A 10MW photovoltaic installation will be online in mid-2009. This will supply power for all other construction activity.

Water management has also been planned in an environmentally sound manner. A solar-powered desalination plant will be used to provide the city's water needs, which are stated to be 60 per cent lower than similarly sized communities. Approximately 80 per cent of the water used will be recycled and wastewater will be reused 'as many times as possible', including for crop irrigation and other purposes.

There are many challenges including: the hot climate and high need for cooling (over 5700 hours per year for cooling and 5200 hours per year needed for dehumidification); dust, which could reduce photovoltaic output; and humidity, which causes the dust to adhere to surfaces. Storage of electricity and thermal energy are also challenges. The personal rapid transit load is highly variable and it is always challenging to manage occupant behaviour (Whittier, 2008).

Lessons from this case study include the following:

- A 100 per cent renewable community requires a large upfront capital investment and planning.
- The phasing is innovative because they are first building a large solar plant to provide power for construction activities.

- The renewable strategy includes both distributed photovoltaic solar systems on homes integrated with power from larger central facilities.
- Infrastructure planning and design is novel. Infrastructure will include storage for heat and electricity. Water infrastructure includes multiple levels of clean water (black, grey, fresh).

Each of these case studies offers some lessons in developing a long-term plan to transition a community toward a 100 per cent renewable community. In the next section, we summarize the lessons learned into a series of steps that a community can take.

A recipe for moving to 100% renewable communities

A community goal needs to be set at the highest level (i.e. the university president, the mayor) but implementation falls to the people 'in the trenches' – the facility, city and utility managers – and must be effectively communicated in order to enlist the support of community members. It is a good first step to set a bold goal and to initiate dialogue. The dialogue should clarify a community's shared values around the importance of using renewable energy whether as a climate-impact mitigation strategy, an economic development strategy or a strategy for energy independence or for other reasons. These reasons will help a community redefine why renewable energy is 'cost effective' for the community by including the external costs of not supporting a renewable energy strategy.[3]

Once there is agreement that 100 per cent renewable energy (or zero energy is a shared value and goal) the next steps might be:

1   **Partnerships and stakeholder roles and relations**
    The paradigm shift required to move to 100 per cent renewables requires a redefinition of roles and responsibilities within a community. Building local relationships early in the process to reach the common objective, regardless of the role each partner played in the past is an important first step. One of the best examples is presented by Reuyl (2006). He states that utilities may well evolve from a producer of a commodity (electricity) into a distributor and manager of services, modelled in part on the banking industry. While banks handle the flow of dollars, utilities handle the flow of electricity. Each industry (banks and utilities) handles multiple sources of deposits (dollars or electricity). Each industry has multiple services to offer (withdrawals/electricity use, deposits/electricity generation, deposits withdrawals/net metering, and various costs of services/different rate structures). Both depend increasingly upon the flow of information to dispense services and maximize reliability and efficiency of the respective industry. The concept of the smart grid being implemented in Boulder, Colorado is very consistent with the utility's evolving role.
    Another example is Toyota, which now has a division in the homebuilding business in Japan, and could leverage their quality brand

name by combining sales of homes and cars (www.toyota.co.jp/en/
more_than_cars/housing/index.html). Likewise a developer might
recognize that the development of a community-scaled utility micro-grid
to interconnect various components of distributed renewable energy with
buildings might be in their business interest.

Going into a dialogue with the focus on the end in mind (100 per cent
renewable) and the reality that roles will need to change to help shift the
paradigm is a good first step. The dialogue can frame tactical plans for
near-term reductions in energy use and shifts to renewable energy with a
long-term strategy for achieving the end goal. Both the tactical and
strategic plans will require refinements. As the community evolves, new
technologies and solutions become available, financing opportunities
arise and the energy landscape changes.

2   **Understand the energy inventory and needs for buildings, vehicles, water
and waste**
In the case studies previously discussed, examples of energy analysis and
inventory are an important early step. There are many tools available to
do the inventory. What becomes more challenging is to quantify the
impact of various technologies in combination with local policy options.
Questions such as what has the larger impact, policy X or Y or policy X
versus incentive A or B, or technology 1 versus policy 2 are challenging
for communities to quantify, rank order and evaluate. Communities
typically rely on a combination of tools and manual analysis to evaluate
trade-offs between policy and technology options, as comprehensive and
unified community energy planning decision support are not generally
available.

Once the energy assessment has been completed, a next step is
developing a rough cut at how much of the goal can be met using
efficiency versus onsite renewables, versus central onsite renewables,
versus offsite renewables. In very general terms, significant reductions are
available from energy efficiency (in existing communities most likely in
the range of 15 per cent to 40 per cent); once these inefficiencies are
exploited, the community needs to find or develop a renewable supply-
side scenario. The tendency of communities is to focus discussion and
planning more on the incremental savings though low-cost and voluntary
efficiency measures and put off or ignore the discussion of the bigger
question – what to do after the incremental improvements are made in the
short term. In an existing community, if the incremental savings from
efficiency are calculated to be in the range of 30 per cent and the
community wants to meet 100 per cent of their energy from renewable
sources, focusing discussion and an implementation plan on only
incremental savings is ignoring 70 per cent of the problem.

Tied to understanding the energy inventory and quantifying the
reduction needed by source, is determining the overall cost and

timeframe. Most likely the costs per community of infrastructure changes will be in the millions or billions of dollars depending on the size of the community. The total cost is an important data point. Once it is known (as shown previously in the Walker Smithsonian example) a finance plan and timeframe for implementation can be determined.

3   **Building strategy**
Goals for new construction and retrofit need to be defined and made measurable over time. One cannot measure what one cannot meter. Without the data, it is hard to know if your community is meeting its target. The building strategy will be a combination of policy requirements and technology options. In terms of technology options, there is no 'silver bullet'. If the stated objective for a community is 100 per cent renewable then a requirement for all new construction should be to maximize energy efficiency in all new construction. (For example, in the US, for most building types a good metric would be to design a new building so that its energy performance is in the order of 50 per cent better than the building code requires by law.) In existing buildings, exploiting energy efficiency is also the most cost-effective strategy. Especially in new construction, all efficiency measures that are lifecycle cost effective should be incorporated.

Once efficiency is exploited, a community with the goal of 100 per cent renewables must install or plan for solar electric photovoltaic systems or solar thermal systems on all available and suitable rooftop space.

US local government and university construction budgets typically include capital costs only, while operating expenses come from another budget line item. This has created a barrier in terms of planning for the added first cost of sustainable new construction. Decisions regarding building construction need to be made based on lifecycle costs (first-cost plus operating). Communities and universities, developers, as well as the banking industry need to work through this somewhat 'artificial' barrier in order to invest in efficiency and renewable measures that are lifecycle cost effective.

The building efficiency strategy must also address changing human behaviour. Universities and other entities are developing many new labs and buildings that use in the range of four to six times the energy of office buildings. In these buildings, strategies to save energy, especially when the buildings are not occupied, are particularly important. As was shown in the NREL example above, human behaviour, including turning off lights and appliances when not in use, is an important efficiency strategy. This is especially important in high-tech buildings on university and research campuses.

4   **The central plant**
The relationship between the community and its utility will be key to decisions regarding how to provide community renewable power from a

central or distributed source, as well as metering. Costs, financing and timeframe surrounding a new renewable power plant are important questions to answer as part of the planning. There are numerous technology options to consider – wind, photovoltaics, landfill gas, hydro, geothermal – that are dependent upon local resource availability. Scenarios for locating renewable technology plants could include siting on land owned by the community, or working with the utility to site and finance a renewable power plant owned by the community at a good wind or solar resource location where the power generated is credited to community and utility.

5   **The transportation strategy**
Understanding the important role that transportation plays in daily life, one must provide options that meet the needs of the individual in a cost-effective and timely manner. For a new community, the most energy saving transportation mode is to provide attractive, safe pathways for pedestrian and human-powered machines. Another energy efficient transportation means is mass transit – with stops at a frequency interval that is timely and low or no cost to the rider. A more energy intensive option is to have specialized lanes for private vehicles that carry multiple passengers (carpools). Since no one fuel (petroleum, natural gas, hydrogen, bio-derived fuels and electricity) will be the only answer, it is anticipated that the impacts of new and/or existing infrastructure along with a global carbon footprint analysis be carefully analysed upfront so that maximum efficiency can be achieved.

Individuals are more likely to choose a fuel that is inexpensive, convenient and perhaps way down the 'environmentally friendly' list. As it stands today, electricity is one fourth the cost of conventional petroleum fuels – and as soon as there are more options for original equipment manufactured (OEM) vehicles such as hybrid-electric and plug-in hybrid-electric versions these will become more popular. Since electricity can be generated renewably – then all planned new development should consider becoming plug-in ready. Neighbourhood electric vehicles (NEVs) provide a low-speed option to conventional vehicles at a fraction of the cost and their use should be encouraged. Electrified bikes are becoming a favourite mode of transportation in China and depending upon the infrastructure and safety aspects during travel, could grow in popularity around the world.

Other fuels, beyond petroleum, could be biomass-derived feedstocks to make – for example – cellulosic ethanol. Early market penetration of corn-derived ethanol is just starting to make its way into the market and much less energy intensive crops beyond corn are being explored for yields three to five times larger than currently exist. Hydrogen and fuel cells, with electric drive, are other options that face greater technical challenges and more infrastructure development than some of the other options mentioned.

The bottom line is that one should encourage efficient and convenient public transportation and reward less vehicle miles travelled to create a more 'local' business network meeting the needs of everyone in very short distances. If private vehicular transportation is essential, then fuel the vehicle with renewable fuels (biomass-derived alternative fuels, hydrogen or electricity from solar, wind etc.).

6   Integrated design

Achieving a 100 per cent renewable community is not purely a technical problem; rather it is a systems integration problem. This chapter presents ideas and approaches, but does not describe actual case studies where 100 per cent has been achieved because none of the examples have been built. There are many systems integration opportunities on multiple levels:

- Energy efficient homes with renewable sources of electricity need to be physically connected via a smart grid to one another and to a central renewable power plant with storage.
- The energy systems for homes and vehicles need to be interconnected and viewed as one system.
- The values in the community need to shift and align with energy efficient behaviour. Institutions within the community, schools, businesses and governments can all model the desired behaviour.
- Traditional roles of stakeholders shift to take advantage of new opportunities. For example, the role of the utility as described in this chapter. A community developer's role might shift to take more responsibility for developing a community's energy infrastructure and the city's role might involve a greater leadership role in advocating clean power.

Research and other action needed to move forward

The key areas of research for supporting the development of 100 per cent renewable communities involve the collection, use and management of information within an integrating framework. First, a standard methodology for baselining direct energy use on a self-consistent, community-wide basis is a critical foundation for developing comprehensive plans. Beyond that, guidelines for estimating and categorizing indirect energy use assist the definition of boundaries and scope for community efforts. The integrated nature of community-based planning requires a similarly integrated approach to assessing and reconciling energy-use inventories for buildings, transportation, central plants, etc. Furthermore, a methodology that structures these inventories according to a standard lexicon or ontology will enable data analysis for inter-community case studies and encourage the development of standard analysis, planning tools and best practice guides for renewable community development. Note that considering greenhouse gas emissions or carbon neutrality in plans, in addition to renewables adoption, further complicates the planning methodology.

A variety of analysis tools already exists for assessing and analysing particular aspects of a community's plan in the short term, but developing comprehensive analysis tools that integrate short-term options with long-term possibilities are needed to allow communities to close the planning gap. The methods for identifying appropriate mixes of near-term, incremental actions and policies do not necessarily apply to the development of strategic plans because the long-term options are coupled to substantial uncertainties in technology, behaviour, markets and policy; thus, it is desirable to formulate long-term plans that explicitly account for these risks and iteratively manage them. Complementary, short-term plans must avoid either closing off or increasing the cost of beneficial long-term possibilities, including financing opportunities. Hence, the loosely coupled development of short-term and long-term plans poses the likelihood of inconsistencies and inefficiencies. Tools soundly based on a methodology that integrates the community-wide and the temporal planning horizons promise lower risk, are more cost effective, and provide more achievable pathways to 100 per cent renewable energy for communities.

As mentioned above, the efficiency or cost effectiveness of a technology does not necessarily guarantee its adoptability, especially if its adoption would require significant behavioural changes within a community. Behavioural economics research into the issues around renewables adoption is critical for identifying adoptable technological pathways and crafting policies that support such adoption. Without a better understanding of the choices available to individuals, businesses and community groups, portions of the long-term plans for 100 per cent renewables could lead to dead ends. The behaviour-economic research needs can be addressed through a combination of theoretical work, modelling and simulation, small-scale experiments, focus groups and case studies.

Finally, the nascent plans and implementations of renewable communities constitute rich sources of potential data that can inform planning methodology, best practice guides, comparative case studies, tool calibration/validation etc. Collecting these data and research results into an information clearinghouse is vital to accelerate the formulation of realistic, integrated approaches to achieve 100 per cent renewable energy adoption in communities. Furthermore, structured repositories of community-level data can be leveraged in regional studies and analyses to provide a wider perspective on progress towards 100 per cent renewables.

Notes

1   There are many competing terms that describe the desired end state of a community or building that meets the majority of its energy needs from renewable energy or carbon-free energy. The terms include: sustainable communities, net-zero communities, carbon-free communities, renewable communities – and in this book, the term used is 100 per cent, referring to the goal of powering 100 per cent of the community needs from renewable energy. There are many nuances in definition, for example, net-zero energy could refer to site energy, source energy or energy costs.

2 If you take the average carbon produced by one person (15t per person at $12/ton) (Tufts Climate Initiative, 2008), the capital offset per person equals $180 per person per year or $540 per year per household of three.

3 In the US, the issue of cost effectiveness has been a barrier to greater use of renewable energy. When comparing the cost of renewable power to other energy sources, the external costs for items such as environmental pollution or transport to remote sites associated with certain energy technologies are not factored in. If these costs were included in pricing energy, the cost comparison between renewable and fossil-based energy would be more accurate. Also, in the US there is ambivalence and polarization of thought around how much more we are willing to pay for clean, renewable sources of power in many communities. This is an important question to build consensus around.

References

ASHRAE (2004) *ANSI/ASHRAE/IESNA Standard 90.1-2004, Energy Standard for Buildings Except Low-Rise Residential Buildings*, American Society of Heating, Refrigerating and Air-Conditioning Engineers, Inc., Atlanta, Georgia

Bailey, J. (2007) *Lessons From The Pioneers: Tackling Global Warming At The Local Level*, Institute for Local Self-Reliance, Minneapolis, MN

Brown, M., Southworth, F. and Sarzynski, A. (2008) 'Shrinking the Carbon Footprint of Metropolitan America', Metropolitan Policy Program at Brookings, Brookings Institute, Washington, DC

City of Boulder (2006) 'Boulder Climate Action Plan;, www.bouldercolorado.gov/index.php?option=com_content&task=view&id=7698&Itemid=2844

City of Boulder (2008) 'Carbon tax', www.ci.boulder.co.us/

Dorsey, J. (2008) 'Fort ZED: Fort Collins Zero Energy District', paper presented to Northern Colorado Renewable Energy Society, 12 February

Kelley, J. (2008) 'Monster homes pit old versus new', *Rocky Mountain News*, 17 May

Lambert, C. (2006) 'The market place of perceptions', *Harvard Magazine*, March–April

Lenzen, M., Wood, R. and Foran, B. (2008) 'Direct versus embodied energy: The need for urban lifestyle transitions', in Droege, P. (ed) *Urban Energy Transition*, Elsevier, London, pp91–120

Little, A. M. (2006) 'The sway of the world', *Grist Environmental News and Commentary*, May

Moezzi, M. (1998) 'The predicament of efficiency', *Proceedings of the American Council for an Energy Efficient Economy (ACEEE)*, vol 4, pp273–285

Poole, R. D. (2008) 'Concept for District implementation of Renewable Power and Energy Efficiency', General Manager, Chief Engineer, Sonoma County Water Agency, January, draft

Reuyl, J. S. (2006) *Sustainable Communities –Business Opportunities for the Electric Utility Industry*, EPRI, Palo Alto, CA

Ritter, B. (2007) *Colorado Climate Action Plan*, State of Colorado, Denver, CO

Sherman, J. D. and Booth Sweeney, L. (2006) 'Understanding public complacency about climate change: Adults' mental models of climate change violate conservation of matter', *Climatic Change*, August

Snider, L. (2008) 'Boulder likely to miss Kyoto Protocol goals', *Boulder Daily Camera*, 3 April

Torcellini, P., Pless, S., Deru, M. National Renewable Energy Laboratory, Crawley, D. and US DOE (2006) *Zero Energy Buildings: A Critical Look at the Definition*, Conference Paper NREL/CP-550-39833, NREL, Golden, CO

Tufts (2008) 'Tufts Climate Initiative. Voluntary Carbon Offset Portal', www.tufts.edu

Walker, A. (2008) 'Renewable Energy Planning: Multiparametric Cost Optimization', presented at SOLAR 2008, American Solar Energy Society (ASES), San Diego, CA, Conference Paper NREL/CP-670-42921

Whittier, J. (2008) 'Masdar Initiative', Brownbag presentation at the National Renewable Energy Laboratory, Golden CO

Xcel Energy (2008) 'Xcel Energy Smart Grid', white paper, document 08-01-311, Xcel Energy, Denver, CO

Chapter Sixteen

Community Life at 100% and Beyond: How to Raise a Renewable Family Without Even Trying

Michael Stöhr

100% renewable – a big step?

Our family lives on renewable energy – 100 per cent. This sounds extraordinary, but there are many individuals and families in the world who are already fully renewable energy supplied – and many may not even be aware of it. Also, our success was made possible only by working closely with many others, and depended on the sustainable urban development principles of our neighbourhood. All we did was to make a slightly larger commitment than others in assembling various readily available cooperative opportunities and to take advantage of the features of our city quarter.

I speak to you from our family's perspective but the story is bigger than that. It tells you about our city quarter, our communal house, parish and various neighbourhood organizations. Key is also the high level of environmental performance applied to the design of our city quarter – and the largely geothermal energy-based DH system implemented by the local government. And our success is testimony to the power of a national legislative framework, providing mandatory payments by utilities to individual renewable power producers – the immensely successful German FiT that has taken so many countries by storm, making it easy for families like ours to succeed almost effortlessly on our path to being fully supported by renewable energy.

A good indication for how unspectacular 100 per cent renewable energy supply can be is that we did not actually realize for almost five years that we had already achieved this seemingly exalted status. Late in 2005, I looked for a piece of information about the German Renewable Energy Act (EEG) on the homepage of Hans-Josef Fell, the Greens parliamentarian and co-author of the German FiT legislation.

I knew Hans-Josef from the Working Group of Bavarian Solar Initiatives. We had worked together in a group of activists to set up Bavarian FiT legislation by referendum before Hans-Josef was elected member of the federal parliament and became the main promoter of the now famous German feed-in law, EEG. His homepage was known as an excellent source of information on that topic. Here, on www.hans-josef-fell.de, I stumbled across his presentation of his personal 100 per cent renewable supply system. When I read his claim that he had achieved this goal, I grew curious. How far along were we ourselves? I sat down, calculated and discovered to my delight that we had exceeded 100 per cent since early 2001. I had been a renewable energy fan since childhood and had done everything I could to make our way of living sustainable, but I had not kept an account of our efforts until that moment.

In early 2006, a colleague made me aware of the German Energy Conservation Competition run as part of the campaign 'Our climate: In search of protection' by the co2online campaign in cooperation with one of the two public television channels, ZDF. I sat down again, calculated more closely and arrived at 113 per cent: 75 per cent in savings and 38 per cent in renewable energy generation. We won the competition in the 'tenant' category, while nine other winners had presented high rates of energy saving and renewable energy use in single-family homes. The jury had appreciated that we had achieved such a high goal through living in an unspectacular multi-family house that was constructed within the financial limits of social housing. We did not have the funds for a house and not even for setting up renewable energy installations. There is no hidden expensive trick that enabled our 100 per cent renewable energy supply – and that makes the difference to almost all other cases. We hope that this will encourage many others to follow our example.

Box 16.1 Key messages

Our household achieved nominal surplus renewable energy generation status simply by lowering energy consumption in electricity, heating and transport, new renewable power purchases to cover remaining demand, plus local renewable electricity generation to help compensate for the remaining fossil components in our energy consumption, in particular, transport. Four ingredients were key to this: choosing the right neighbourhood to live in; using readily available community-organizing techniques; picking cooperative models of solar power asset ownership and generation; and having available the choice of true, new renewable electricity purchasing arrangements. Our next hope would be to eliminate the use of oil in transport altogether by our car-sharing organization switching to solar electric powered vehicles and public transport switching to fully renewable energy supply – but that is not within the reach of our decision-making powers.

Cover the first base, the choice of neighbourhood: Communal support and sustainable urban development are critical

Messestadt Riem at Munich's eastern fringe is a model for a post-fossil, post-'airtropolis' redevelopment: it occupies former Munich airport, although the airport simply expanded on a new site to the city's north. The small city centre was conceived as a showcase for urban sustainability, using the state of the art at the time. Ultimately, 16,000 people are expected to live in the area. Unlike much of the rest of Munich, Messestadt Riem is home to lower than average income groups and a higher than average number of children: precisely the sort of population usually considered to be too deprived of income or means to afford a 'sustainable' lifestyle.

The DH system is fed by a geothermal power station using heat from the Upper-Bavarian hydrothermal stratum found 2700m below our neighbourhood. Eighty-five per cent of the heat consumed in the quarter comes from geothermal energy and the balance is supplied by a natural gas-fed CHP station and a natural gas peak electricity generator.

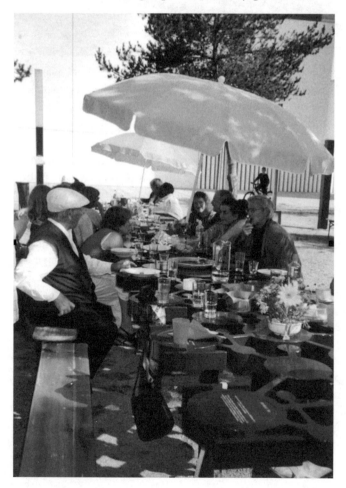

Figure 16.1 Residents of our quarter sit down for a meal at the Platz der Menschenrechte (Square of Human Rights): On each plate, an article of the Universal Declaration of Human Rights is engraved

Source:
Helga Rätze-Scheffer

Figure 16.2 View of our city quarter from the top of our house with SOLNA PV plant in front, and ecumenical church and shopping centre Riem-Arcaden in the background

Source: Michael Wippermann

Messestadt Riem was designed as a compact yet green urban quarter of short walking distances and a high level of integration into the wider public transport network. Individual car traffic is dramatically lowered and a number of car-limited house communities are set up, among them the one where we live. The inhabitants of these house communities declare in their agreements that they will not own an individual car. Exceptions are possible, but the number of cars per house is limited. As a result, the number of cars per person is half of the area average, with the effect that traffic noise is non-existent and there is an exceptional degree of safety for children playing in the streets. Instead of owning an individual car, many inhabitants of these house communities are members of one of the two car-sharing organizations in Munich.

The urban design has three consequences for the energy balance of all area inhabitants: the compactness of the buildings reduces heat demand, the high level of public transport services and the car-sharing option lowers overall transport energy and transport infrastructure (i.e. parking lots) demand, and the DH network fed largely by geothermal energy ensures a high level of heat supply from a renewable energy source.

Also helpful: Choose a cooperative living arrangement

The only way an average family with limited means can afford full renewable energy supply is to organize and build on a network of wider community support. There is no magic in the number 100 per cent, and neither does it require iron discipline and perfection to achieve it. We exceeded 100 per cent renewable energy supply despite the numerous compromises we had to make. The keys to a renewable life are new or revived forms of networking and cooperation. This message will become clearer when I specify our energy savings and renewable energy measures. The community organization was essential in achieving 100 per cent renewable energy supply. WOGENO e.G., the German acronym for housing cooperative, is one of several established in Germany over the last two decades dedicated to social and environmental self-governance. WOGENO was founded in 1993, and we joined in 1996 with a group of people planning the construction of the house we live in. Cooperatives are an alternative to owning and renting, a hybrid solution. Cooperative members are owners, because they acquire a certain number of shares, depending on apartment size and level of income. On average, shares corresponding to about 10 per cent of the value of an apartment are held. A member is also a tenant: a monthly utilization charge is due, the equivalent of a rent. A cooperative apartment cannot be owned or sold, but the right to live in an apartment can be bequeathed. This is a constitutional principle of our housing cooperative whose main aim is to provide affordable living spaces for all members of the community. In order to achieve this, living space must be excluded from speculation and profit making. Housing rights are attributed by an elected committee in charge of judging the priority order of candidates for a specific apartment or transferred from occupant to occupant through bequeathing only, since this does not boost a speculative inflation of asset value or rents.

Life in a WOGENO house was the foundation for our most important energy saving and renewable power successes. We would not have been able to afford a house in Munich, and a standard rental unit would not have afforded the necessary degree of control over one's energy future.

What do we mean by 100%?

We combine energy savings and use renewable energies, and conservation is the larger part in this equation. We consume 75.2 per cent less electricity, heat and transport energy than an average German family of the same size. And the renewable energy generation component in our family reduces our reliance on fossil fuels by another 38.4 per cent of that notional average family consumption level. The total adds up to our 100 per cent-plus performance. We are also very conscious about reducing the embodied energy in our lifestyle – but it is not included in this calculation. By 'embodied' we mean the energy that was required to satisfy our daily consumption of goods and services: food or clothing, for example, or what has been required to manufacture, transport

Figure 16.3 Our cooperative-owned multi-family apartment house consisting of two buildings connected by a glass-covered bridge at the level of the second floor

Source: Michael Wippermann

Figure 16.4 Inhabitants contributing labour to house construction (construction of bicycle sheds)

Source: Helga Rätze-Scheffer

and assemble the materials and components our building or cars are made up of. More on that point later.

Our energy consumption and supply profiles do not match entirely. We account for 4.66 times less than the reference family's non-renewable energy consumption level through saving measures and renewable energy purchase/generation in the sector of electricity. Energy needs for room heating are completely avoided, i.e. 100 per cent of non-renewable energy is being abated in this sector. Energy for hot water is not conserved well, but 85 per cent of it is renewably supplied. Comparable transport energy use is reasonably well reduced at 50 per cent of the average wider area household, the rest is still fossil-fuel based. As a last measure we offset our transport energy component through surplus generation of renewable energy.

The offsetting of non-renewable transport energy through the surplus production of renewable electricity matches the expected future development of energy supply well: renewable energy generation from wind and photovoltaic installations is developing so quickly that there will be surplus electricity production at least from time to time and at local spots with many generation facilities. This surplus production can be stored by a fleet of electric vehicles. Given inexorably rising oil prices and a debate about the limits and (non-)sustainability of biofuels this will be the only way to ensure individual mobility in the mid term and long term.

Table 16.1 Energy consumption by our family in comparison to the German reference family

	Electricity	Room heating	Hot water	Transport energy	Total
Reference family	5300	35,000	4000	20,000	64,300
Stöhr household	1800	0	3330	10,837	15,967
Energy saving	66%	100%	16.8%	45.8%	75.2%
Rest covered by renewable energies	1800	0	2831	663	5293
Rest covered by fossil energies	0	0	500	10,175	10,674
Part of renewables in purchased energy	100%	–	85%	6.1%	33.1%
Non-renewable energy abatement through saving + renewable energy purchase	100%	100%	87.5%	49.1%	83.4%
Generation of renewable energy in addition to own consumption	19,405	–	–	–	19,405
Contribution of renewable energies to non-renewable energy abatement	400.1%	0.0%	70.8%	3.3%	38.4%
Contribution of energy saving + renewable energies to non-renewable energy abatement	466.1%	100%	87.5%	49.1%	113.6%
Contribution of each sector to non-renewable energy abatement	38.4%	54.4%	5.4%	15.3%	113.6%

Note: The values in the first indicate the mainly non-renewable energy to be abated. All values in kWh. Percentages relate to a German reference family of the same size, whose consumption pattern matches the German average for households for electricity and heat, and for individuals in the transport sector.

Steps to 100% renewable electricity

Before engaging in the purchase and generation of electricity from renewable sources, we focus on savings. We consume 66 per cent less electricity than a German reference family.

The most simple ways of saving electricity are measures known for several decades and yet, amazingly, are not yet widely adopted even in seemingly green Germany: use of supremely efficient household appliances, cut-off of stand-by electricity consumption by a connection plug board with a switch, and general awareness and behavioural adjustment to the need to minimize electricity use. Though efficiency is generally seen as a virtue, it is connected mentally to loss of quality of life and thus its practical implementation is lagging far behind what could be possible.

We do not know exactly why we save so much electricity compared to other German families, but one key to achieving high energy savings in cities lies in life in multi-family buildings, allowing not only the electricity consumption of services but also the distribution of hot water to be shared by 28 households. Another key to savings is replacing electricity by hot water whenever possible. So, our dishwasher is connected to the hot water tap. This make an enormous difference since pre-heated water reduces or avoids the need to heat water for dish-washing electrically. Food is often prepared in a wall-mounted steam cooker, also connected to the hot water tap. Compared to cooking a pot of water, much less water needs to be heated up and the temperature difference that is to be overcome is much smaller.

Switching to 100 per cent renewable electricity supply is the easiest step of all. In Germany, one only needs to choose a supplier who sells electricity that is 100 per cent renewable. We have chosen a supplier that acts as an agency for selling farm-produced renewable electricity.

It ensures that all our electricity comes from new renewable energy plants, and thus that renewable electricity production actually increases.

Electricity saving and purchase of the remaining electricity that we consume from a 100 per cent renewable electricity supplier brings our non-renewable electricity abatement rate up to 100 per cent.

Surplus renewable electricity generation for off-setting fossil transport energy

Citizen-owned photovoltaic installations set up with our neighbours

In addition to purchasing 100 per cent renewable electricity from generation facilities owned by farmers, we are also owners of renewable electricity generation plants, via shares in two photovoltaic installations and a small share of a windpark. All three installations are citizen-owned community installations. While the windpark was implemented by a developer, we have organized the installation of the two photovoltaic plants ourselves.

The first installation (SOLNA) was set up on the flat parts of the two building parts that form our house; the second (SOLKIZ) was installed on three roofs of the ecumenical church centre in our city quarter.

Figure 16.5 Members of SOLNA during a break off work installing a PV plant

Figure 16.6 Display panel showing the yield of the SOLNA PV plant in the entrance area of our house

Source: Michael Stöhr

Table 16.2 Breakdown of our share of the electricity generation of different plants compared to our energy/electricity consumption and the electricity consumption of a German reference family

	electricity generation (kWh)	in % of our total energy consumption	in % of electricity consumption of reference family	in % of our electricity consumption
PV plant SOLNA	1900	3.0%	35.8%	105.6%
PV plant SOLKIZ	838	1.3%	15.8%	46.6%
wind park share	16,667	25.9%	314.5%	925.9%
	19,405	30.2%	366.1%	1,078.0%

The SOLNA: 'Solar Neighbours Riem' private company

For the photovoltaic plant on our apartment building, 23 neighbours founded the operation company SOLNA – Solare Nachbarn Riem GbR – the solar neighbours Riem personal company – in 2000. SOLNA has purchased the photovoltaic plant, operates and sells the electricity to the local electric utility that pays the mandatory minimum FiT for solar electricity. The shares signed by SOLNA members range from €250–2500. The small minimum share was set to also allow inhabitants with little income to be a member of SOLNA. Jointly, about €13,500 of equity (some $18,000 at the late 2008 exchange rate) was put together and the balance was financed by a small lump sum subsidy of the City of Munich and an interest-free loan from Kreditanstalt für Wiederaufbau (KfW) (German Bank for Reconstruction).

The inhabitants of our house did not only participate financially, but contributed also with personal labour to the realization of the photovoltaic plant, providing 'sweat equity'. The plant is operated on a voluntary basis.

We had to make a seemingly trivial but fundamental decision concerning the position of the photovoltaic panels on our house, illustrating the cooperative nature of our work. Should we put them on the large area of the two roofs that are slightly tilted to the east, thus accepting a lower yield due to non-optimum orientation and tilt, or should we put them on the flat parts of the roofs, thus reducing considerably the roof area on which we could hold gatherings, but ensuring a higher yield with well-oriented and tilt panels, only reduced slightly by shading due to the higher rising parts of the roof system on the eastern side? Finally, we decided on the latter. The flat parts of our roofs became smaller terraces that are now very quiet places for those who seek a retreat from the generally very active life in our house.

The SOLKIZ: Solares Kirchenzentrum Messestadt Riem GbR

The SOLKIZ photovoltaic plant was constructed in 2005 on the top of three roofs of the new ecumenical parish centre in the Messestadt Riem, Munich, which was finished in that year. It is almost three times larger than the SOLNA photovoltaic plant, but has been set up exactly in pattern with SOLNA. Again, we have founded a personal company, the SOLKIZ – Solares Kirchenzentrum Messestadt Riem GbR (Solar Church Centre Messestadt Riem Personal Company). The contractual and financial setup was similar to that of the SOLNA plant.

Figure 16.7 Ecumenical church centre with SOLKIZ photovoltaic plant seen from the church tower

Source: Gerhard Endres
Note: Photovoltaic panels are on three different roofs – left behind, centre and at the right out of picture

Within the first three years, the SOLKIZ photovoltaic plant has produced 47,400kWh, corresponding to 922kWh/kWp, a value that is higher than for SOLNA but below the average yield of new photovoltaic plants in Munich. Here, the reason is that the architect of the ecumenical church centre in the Messestadt Riem requested that the photovoltaic panel should not be visible from the ground (now also the reason why it is so difficult to provide good photos of the installation) thus obliging us to keep a tilt angle of 5° on one roof and 17° on two other roofs. This is much less than the optimum tilt angle at the latitude of Munich, which is 28°. As a result, we yield less solar energy in winter and a bit more in summer than a plant does that keeps the optimum orientation and tilt angle. Over the year, we lose about 10 per cent of the solar yield compared to an optimum plant. Here, old-fashioned ideas about the incompatibility of photovoltaic panels and highly aesthetical architecture (and the fact that the two churches did not request from the architect right from the outset to integrate photovoltaic panels as an element of the church centre's building shell) compromised environmental efficiency.

The most important contribution: Bringing heat demand down to zero

What surprises most people is that we do not need to heat any more. As mentioned above, we live in a building cooperative house with 28 households constructed in the cost limits of social housing, i.e. there is really no expensive technology put in it. The building fits the energy consumption limits of a low-energy house. However, the two parts of the building are large and almost cube-shaped, i.e. have a very good ratio of surface to volume. This means that the low-energy standard that is by definition a relative standard that depends on the surface/volume ratio of the building, corresponds to a very low absolute heat demand of only 36kWh per square metre a year. This is a direct consequence of the compactness of our city quarter, which obliges builder-owners to construct very compact houses. In comparison, single-family houses can be declared as low-energy houses though the heat demand might be as high as 100kWh per square meter a year.

In the planning phase, the sanitary and heating engineer worked out that we could lower the heat demand of the building even further by installing controlled ventilation with heat recovery in all apartments. This would have allowed us to forgo heat radiators in all apartments. In this case, only a small peak load heating system would have been needed that could have been a heat pump integrated into the heat recovery of the controlled ventilation. However, when this option was presented in the planning meeting of the future inhabitants, several persons objected firmly to this idea. Some had had bad experiences with air conditioning systems in offices and were afraid that something similar was on offer. Hence, no consensus on the controlled ventilation with heat recovery was reached, and a floor heating system was installed instead. Here, the democratic principles of WOGENO compromised the environmental efficiency that could have been achieved.

Nevertheless, WOGENO chose to provide the option for individual families to install a controlled ventilation system with heat recovery, which is normally conceived for refurbishment of existing buildings. This system is to be installed in each room and consists of an opening (about four inches wide) through the wall, which contains a tube with the heat recovering device and a small fan (as used in personal computers) expelling and taking in air at 80 second intervals. Although the costs per room were steep (€1000 or $1340) we chose that option for all five rooms in our 100m² apartment and thus reduced heating demand down to zero.

Thirty-six kWh/qm/yr is a theoretical value. The actual heat consumption very much depends on individual heating behaviour. Actual heat consumption – not considering us who do not heat at all – ranged from about 16–160kWh/m²/year before we started a campaign on energy saving behaviour in our house, i.e. varied by a factor of ten. The main difference is in the ventilation behaviour of the inhabitants. People who do have intermittent ventilation consume less than those who keep their windows always open while the heat radiator is turned up to maximum. Contrary to some views, there is no appreciable difference in indoor air quality.

We take showers as frequently as everybody does!

We may not heat our apartment by other than inherent and solar heat sources (body heat, waste heat from electric appliances and direct solar radiation through the windows), but we do not renounce the copious use of hot water. We do take showers as often as 'normal' people do! In fact, we do not save hot water use when compared to other households, despite hot water saving water taps that are installed as a standard component of all bathrooms and kitchens in our house. A main reason for this is that we use hot water to lower electricity consumption: our dishwasher and the wall-mounted hot steam cooker are directly connected to the hot water tap. But we also don't waste water either.

However, as our house is connected to the DH network of our city quarter Messestadt Riem, and the energy for providing our hot water is generated by the mix of that heating network: 85 per cent geothermal energy and 15 per cent natural gas, our hot water supply is predominantly based on renewable energy. Again, this is not to our credit but is something we owe to the sustainable urban development of the Messestadt Riem, which was decided by the city council and the DH network that was built and is operated by the municipal utilities.

Sustainable transport mix

In the transport energy sector, we consume about half the energy for our private mobility needs of the German reference family. Here, we profit first of all from the fact that we live in a large city with a well-developed public urban transport network. In particular, the sustainable urban development of the Messestadt Riem quarter has created short distances to schools, recreation facilities, shopping etc.

Our average annual mobility pattern is shown in Table 16.3. First of all, it is worth pointing out that we limit private air travel for the whole family to only once in five years. This was the typical average for the last 10–15 years and the usual destination was Southern Europe. Note that this air travel once in five years accounts for about one third of our transport energy consumption!

Most of our mobility demand is met by public urban transport – essentially underground train for me going to the office – and railways. Only a small part is met by cars. Yes, cars, not a car. As mentioned above, we have some 250 cars at our disposal, being members of the car-sharing organization STATTAUTO München, offering transport vans and small buses as well as a range of passenger cars including mini, small and family cars. We book a car only three or four times a year as the rest of our travel needs can easily be met by other means of transport.

Table 16.3 Breakdown of our average annual mobility and related energy consumption

	public urban transport	railway	car	airplane	sum
person-km/yr	13,125	20,000	1100	4000	38,225
kWh/km	0.2	0.2	0.6	0.89	0.28
KW/yr	2625	4000	660	3552	10,837
% of renewable energy	10%	10%	0%	0%	6%

Source: Michael Stöhr

Difficult to calculate but very important: Embodied energy

There is a part of our energy consumption that we could not calculate and compare to a German reference family: embodied energy. This means energy that is needed for the production of goods that we use, for instance, for the installation of our house. A modern low-energy house consumes about one third of its life-cycle energy need during the construction phase alone. This energy is needed for producing the construction materials and for the construction itself. There are two ways of minimizing this energy: first, reducing the effective living space per person and, second, proper choice of construction materials.

The WOGENO cooperative has set limits on the apartment area per person and balances this by providing a number of collective facilities in its houses. This comprises normally a community room for gatherings, a guest apartment and one or more community rooms in the cellar. In our house, the community rooms include two workshops in the basement floor, one for wood and metal working and painting, another one for tailoring and finer works, as well as a play room for children. In addition to that, we have set up community facilities such as sheds for bicycles and gardening, a common playground between both houses, next to our community room and the terrace in front of it, and the major part of the garden is communal, including vegetable patches that are distributed to those inhabitants who want to grow vegetables.

The effect of the community facilities is that the effective living area per person is reduced. This has two consequences for the energy need: first, the area to be heated is reduced, thus reducing the heat energy demand, and second, the embodied energy in the building is reduced!

The second way of reducing embodied energy in buildings, the choice of construction materials with low energy embodiment, was only partially followed in the case of our house. Energy saving steps included limiting the use of bricks (450kWh/m³ embodied energy) and concrete (250kWh/m³) and making much use of wood (5kWh/m³).

Closely related to the community facilities is the sharing of goods or services that are not constantly needed and hence can easily be shared – provided that people take care of community property, a prerequisite that requires a well-functioning house community. Sharing of goods saves production energy for these goods.

We save the major part of embodied energy by our limited use of cars: STAT-TAUTO München keeps only 20 cars per member, thus cutting down the production energy for cars by a larger factor (not exactly 20, because the cars are run only for about a year, before being resold). The production of cars, and also of parking lots, streets, etc., consumes a lot of energy and we cut down this production energy – in exactly the same way as most of our neighbours do!

A further area where we save embodied energy is food. Almost all our food comes from ecological production and most of is comes out of the region of Munich. Further, we consume little meat, though we are not entirely vegetarian. Hence, we get positive indicators for all factors that play a role in the energy consumption of food production: eco-agriculture saves energy embodied in mineral fertilizers and plant protection chemicals; regional food sources saves transport energy; a low level of meat consumption saves energy because feeding plants to animals before using animal products for human nutrition multiplies the energy requirement; and seasonally grown food in open fields needs less energy than food grown in glass houses. As a result, our embodied energy consumption for food is far lower than the German average – as it is for most of our neighbours who have similar consumption patterns.

The cooperation and networks that allow us to achieve these positive indicators are among others: Tagwerk GmbH, a company for the regional marketing of eco-food delivered weekly to the households, UNSER LAND, a regional food marketing network and company for agricultural products from the region, selling mainly eco-food, and a local eco-food shop.

Finally: Quality of life!

Well, finally, I do not want to say in words that living with 100 per cent renewable energy goes hand in hand with a high quality of life. Instead I will let the pictures of life in our house and in our city quarter speak for themselves!

Websites

Building cooperative WOGENO: www.wogeno.de

City quarter Messestadt Riem: www.messestadt-riem.de

Pictures of Messestadt Riem: www.endres-bildung.de

100 per cent renewable electricity: www.naturstrom.ag

Our windpark: www.windpark-saar.de

Solares Kirchenzentrum Messestadt Riem – SOLKIZ GbR: www.sankt-florian.org

Geothermal heat in Messestadt Riem: www.swm.de

Car sharing: www.stattauto-muenchen.de

Organic food from the area: www.unserland.info and www.tagwerk.net

German Climate Protection Campaign: www.c02online.de

Figure 16.8 A typical scene in the Messestadt Riem: The 'Promenadenfest' (promenade festival) where everybody contributes to the buffet and everything is shared

Source: Michael Stöhr

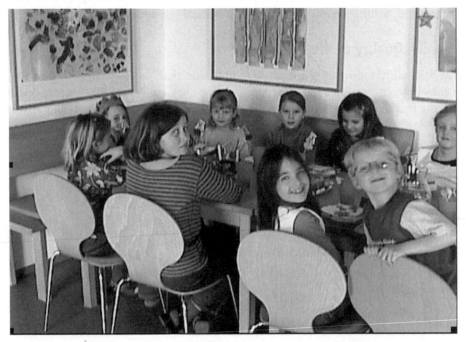

Figure 16.9 A child's birthday party in our community room

Source: Michael Stöhr

Figure 16.10 The terrace and play ground extend the community room outdoors: Here, many gatherings and parties take place

Source: Michael Stöhr

Figure 16.11 The annual cooking party in the community room of our house has become a tradition: Each participant prepares a part of the menu and all taste it together

Source: Helga Rätze-Scheffer

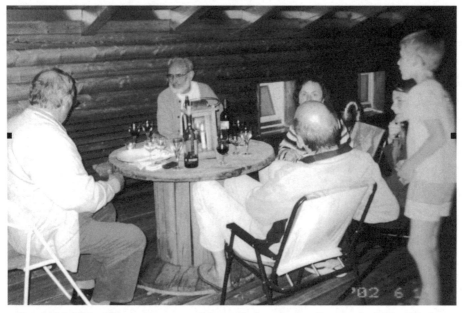

Figure 16.12 Another party on top of the roof of our house: Despite the PV installation, there is enough space left to gather

Source: Helga Rätze-Scheffer

Figure 16.13 Cooking with the sun: Barbara Fredmüller prepares food with her solar cooker

Source: Michael Stöhr

Chapter Seventeen

100% Renewable Life: One Man's Journey for a Solar World

Martin Vosseler

Crossing the Atlantic with 100% solar energy

Imagine – waking up in a narrow hull of a solar catamaran, in a cove of the Rhine River near the German/Dutch border, on a foggy October morning. The air is filled with bird voices. When I crawl on deck I see thousands of wild geese, taking off in groups for their flight south, flying in V-shape formation. Or imagine a clear night on the Atlantic. The dog star is so bright that its light is reflected like a golden ribbon on the dark velvet of the ocean. Suddenly a splashing sound – a dolphin's back emerges. The animal is dancing around our boat. Where it passes fluorescent algae are lighting up – microgalaxy floating in the dark waters.

What a wonderful unique planet we are living on: the perfectly shaped ball is exactly the right distance from the sun so we don't burn and don't freeze. It is exactly the right size so the water of the sea, the rivers and the lakes doesn't disperse into the universe. It has this awesome 'skin' of air – four to five miles of air with oxygen – that allows us to breathe and to live – a very thin layer, a distance that can be walked in less than two hours; and the sun, a powerful nuclear fusion reactor safe distance to Earth, sends us all the energy that we need.

During our Atlantic crossing on the solar catamaran, Sun21, we become aware of the multitude of miracles that make this planet Earth possible. We also realize how vulnerable this unique life system is. We witness sea pollution and the depletion of the maritime fauna; but we learn at the same time how well the combination of renewable energy and energy efficiency works!

What does it take to cross the Atlantic with the energy of a hairdryer or an iron? Six Swiss dreams that come together.

Mark Wuest, solar boat constructor for more than 20 years, dreams of crossing the Atlantic with such a catamaran. David Senn, marine biologist, dreams of making an Atlantic 'Transsect' – to examine a plankton sample every day during an Atlantic crossing. Beat von Scarpatetti, historian and

Figure 17.1 The solar catamaran Sun21 in Miami, Florida

Source: Dylan Cross

founder of the 'Swiss Club of Carfree People', dreams of putting his feet on US ground without a drop of oil. Michel Thonney has crossed the Atlantic already several times on sailboats. He dreams of doing it again, applying his navigation skills to a pioneering adventure. Daniela Schlettwein, a Swiss medical doctor, uses her financial resources to support ecological projects. She dreams of making an unusual solar project possible, after decades of promoting renewable energy and energy efficiency. I prepare for my SunWalk 2008 – walking through the US for the promotion of 100 per cent renewable energy. I have the dream to cross the Atlantic in a sustainable way and, at the same time, to promote 100 per cent renewable energy with this adventure.

Sometimes things fall into place very quickly when dreams come together. In December 2005 we decide to go ahead with the project. On 16 October 2006 the then Swiss President, Mrs Micheline Calmy-Rey, christens the boat in Basel – and off it goes starting for its seven-month journey to New York City. Basel – Rotterdam – Seville – Canary Islands – Martinique – Miami – New York City, where we arrive on 8 May 2007. The actual crossing from the Canary Islands to Martinique takes 29 days.

Box 17.1 Sun21

Solar catamaran Sun21: 46 feet long, 22 feet on the beam. Draft: 3 feet. Weight: 12 tons. Two 'LEMCO' electro motors, 8 kW each, with an efficiency of 90 per cent. The solar energy is harvested by two 5kW modules (about 65m²) that are located on a roof installation. We have 0.8 tons of lead acid batteries in each hull, 48V DC. The propellers are made out of carbon. Maximum speed: about 9 knots (about 16.5km/hour). Constant speed: about 5 knots (about 9km/hour) 24 hours a day. There is cabin space for five people, a kitchen and a bathroom.

We can demonstrate with this journey that if we come down from our high energy waste level, if we combine it with high energy efficiency, renewable energy is sufficient for our energy needs. We have travelled from Europe to the US with an average of 1700W, the power that a hairdryer or an iron needs – a 12t boat, five adult men, five computers, all instruments, a refrigerator, ten cabin lights, four position lights, a motor for the anchor and a satellite phone, all travelled thanks to the 90 per cent efficiency of the two electro motors that propelled the boat. Travelling on this boat is very comfortable – no noise, almost no vibrations and no exhaust fumes; therefore the dolphins like to visit us and we can sleep on the boards that cover the motor.

Towards 100% renewable energy: Step by step

On 1 April 1975, the construction machines arrive in Kaiseraugst – 9 miles from the centre of Basel, Switzerland – for the construction of a nuclear power plant. Hundreds, later thousands of people – from all age groups, professions and political parties – become involved in the non-violent resistance against this project. The territory was occupied. The construction was stopped. In 1988, the project was abandoned.

Kaiseraugst was the beginning of my involvement with clean energy. Later, as a research fellow at Harvard Medical School in Boston, I was a student of Bernard Lown, the founder of IPPNW (International Physicians for the Prevention of Nuclear War, who were awarded the Nobel Peace Prize 1985). I learned about the connection between atomic weapons and nuclear power and started the Swiss chapter of IPPNW in 1981. We invited David Freeman to Switzerland. He started a successful energy efficiency programme as the CEO of the Tennessee Valley Authority and cancelled several nuclear power plant projects.

I tried also to make my life more energy efficient. I stopped driving. I disconnected my 16th century house in the mountains from the electric grid for two years and lived with the old wood stoves and with candles. Later I made good roof insulation and installed a solar water heater and a heat pump in the mountain house, and a solar water heater and wood pellet heating in a small wood house that I let to people. I bought solar electricity for my houses. I became a vegetarian 20 years ago. I stopped flying. But the biggest joy I discovered was walking.

The joy of walking

In 1999 I walked from Konstanz, Germany, to Santiago de Compostela, Spain; in 2003 from Basel, Switzerland, to Jerusalem. The motto was 'There is Enough Sun for All of Us'. I walked through the US, from Los Angeles to Boston.

Walking is a very simple way of travelling with renewable energy fuelled by food. I experience it also as a prayer with body and soul, connecting with Mother Earth, step by step. I discover how healthy walking is – if everybody walked one to two hours a day, didn't smoke, only drank moderately and ate

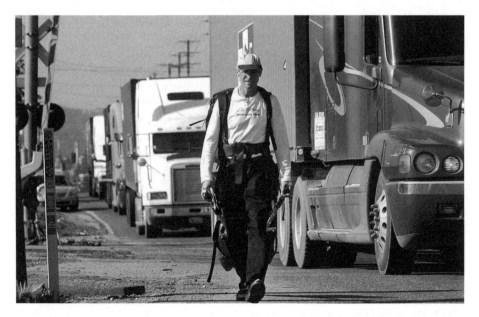

Figure 17.2 Sunwalk 2008: Martin Vosseler in Los Angeles on 2 January 2008

Source: Brad Graverson, Daily Breeze

a healthy diet, we could close a considerable part of our hospitals. I experience that what is good for me is good for the planet. When I walk all my senses are open: I see so many miracles of nature. After some weeks of walking I experience the curved shape of our planet and enjoy the feeling of walking on this amazing life ball. During my walks I live very much in the 'here and now'. The days are filled with surprises – I never know what landscape will appear behind the next bend, where I will find food, where I will stay for the night and whom I will meet. On my walks I meet many helpful, hospitable people. I encounter them at a personal level; and there are many opportunities to exchange opinions, ideas and visions about our energy future.

I think of the lady owner, with red dyed hair, in a purple robe, in a full restaurant of a small town in central France. 'Une table pour le pèlerin!' ('a table for the pilgrim!') she shouts into the room. They bring a small table and they serve a delicious four-course vegetarian dinner, 'on the house'. And after the meal she addresses the whole crowd: 'I am not only a restaurant owner, you know, but also an opera singer; and I will sing a song for this pilgrim'. And she sings with her wonderful voice: 'Pèlerin, suit mon chemin!' ('Pilgrim, follow my path!').

I recall Ahmed and Sanae in Saraýçik in the Turkish mountains. I arrive at dusk. It's cold with a flurry of snow. I ask a shepherd where the Muhtar, the town president, lives. 'There, in the house with the green roof.' I knock at the door: 'I am on a pilgrimage to Jerusalem. Do you know a place where I could stay for the night?' Ahmed invites me into his house. Sanae brings a table cloth and spreads it on the floor. She brings a variety of delicious dishes. Then she prepares their sleeping room for me. She makes a big fire in the stove and prepares the bed. And soon I am dreaming in the warm bed listening to the singing stove.

It's a very hot day in Virginia. In Lacey Springs, Conny parks her car in front of me. She brings me a big cup of cool spring water and a plastic bag with ice and a towel. From then on I wipe my face every half an hour with the icy water.

100% renewable energy and energy efficiency: Number 1 priority

As a doctor I face situations where only one action is needed, individually and globally, for example, if a patient suffers from an arterial bleed there is only one thing that has to be done immediately – stop the bleeding, 100 per cent. If somebody has a cardiac arrest the life saving measure is to restore a heart rhythm that guarantees a sufficient blood circulation – without any delay.

I see climate change as a life threatening process at the global level – the symptoms include extreme weather conditions, rising sea level, reduction of the Gulf Stream circulation. A climate collapse may become possible. There is no cure for such a serious condition; therefore all efforts have to be invested into prevention. Burning fossil fuels contributes to global warming; this kind of energy production is not compatible with the Earth's life system. The same is true for nuclear power – enormous risk potential and radioactive waste. The 'medical prescription' is clear: 100 per cent renewable energy and energy efficiency.

From centralized energy production to a global energy democracy

We are living in a transition time. The biggest part of energy production today is still controlled by a relatively small number of big corporations. The future will be different. In a decentralized energy democracy each citizen can become an energy producer.

The following argument can still be heard often: 'it will be difficult to replace a big part of fossil fuels and nuclear energy by "alternative" energy'. 'Alternative' energy – is it really alternative? Without 'alternative' energy it would remain dark in the morning. The temperature on this planet would be minus 240°C. There would be no precipitation, no water and no food. There would be no life, no animals, no plants and no human beings. Earth would be a cold, black, dead planet.

'Alternative energy'? No, main energy! Basic energy! Renewable energy! Solar energy! Without this main energy nothing would live, grow, move. Next to 100 per cent of the Earth's energy that we use is solar energy. It provides the conditions for life, for food, for water, for growth and movement. The so-called 'conventional' energy forms such as oil, gas, coal and nuclear power provide less than 1 per cent of the Earth's energy needs. In less than an hour

Figure 17.3 Albuquerque, New Mexico, 1 March 2008

Source: John Schaefer

the sun sends the energy onto the planet – or in one year onto 10 per cent of the Sahara's surface – that corresponds to the world's total power consumption.

What can we do? Sixty per cent of the so-called 'conventional' energy we use today is wasted. Optimal energy efficiency can replace these 60 per cent already. The remaining 40 per cent can be substituted by renewable energy that is available in abundance – if we are determined to do that. We have all the technical means to harvest renewable energy. This change of our energy system is a huge opportunity for the world's economy as well. Millions of jobs will be created – jobs that make sense.

The new energy democracy needs the cooperation of all people on this planet and the political framework that favours renewable energy and energy efficiency. Not everybody can walk or go to sea for several months at a time. But everybody can figure out what individual steps they can make possible; after having read an interview about my SunWalk in the *Navajo Time*, a lady in a supermarket in Window Rock, NM, recognized me. She tells me: 'I read about your SunWalk. I told myself: "if this 59 year old man can walk 3700 miles from LA to Boston, I can also walk from my home to the supermarket." So this morning I walked for the first time to work and actually, I enjoyed it very much.'

More and more people believe in the transition to a 100 per cent renewable energy future and help to prepare it with their own steps. Together, with awe and enthusiasm for our miraculous planet, we will make it.

Index

biofuels 21, 22, 23, 42, 98, 101, 105,
121–122, 205, 284
cooking oil as 121, 122
emissions from 124
ethanol 112, 113–114, 284
and food production 69, 252
pan-European project for (RENEW)
110
promotion campaigns 114, 125
biogas 12, 39, 72, 96, 98, 116–117, 123,
129, 177
challenges with 118
cooperative production of 121
costs 118
and emissions reduction 136
and fertilizers 136
gasification process 107, 108–109
household production 133–137
and sanitation 133, 136
subsidies 135
biomass energy 30, 31, 33, 37, 71, 98,
105, 142, 258–259, 277
costs 50
exports 120
land used for 254, 255, 257
in networks 71, 72, 78, 79, 107
pollution from 130
see also straw-based heating
biomethanization plants 127
bioplastics 69
black liquor 127
blackouts 22, 61
boat, solar powered (Sun21) 307–309
Boise (Idaho, US) 18
Bonn (Germany) 55
Booth Sweeney, L. 269–270
Boston (Massachusetts, US) 88
Boulder (Colorado, US) 213, 271–274
Climate Action Plan 272
smart grid initiative 272–274, 281
Brandt, Willy 52
Britain 169, 231, 232
distributed generation in 31
wind power in 28, 66, 217
zero carbon homes policy in 12
see also London; Woking
Building Energy Management System 152
building-integrated agriculture 45, 147,
229–240
benefits to buildings of 233
carbon emissions of 233

constraints on 238–239
cooling systems in 232–233
and education 235
and employment 234
energy saving in 229, 232–233, 237
environmental benefits of 233
future of 239
horizontal rooftop greenhouses
234–235
hydroponic see hydroponics
need for 229, 230
and public health 231, 233–234
PV systems in 233
vertically integrated greenhouses
(VIGs) 236–238
buildings, low energy 87, 89–90, 91, 104,
128, 139, 145–150, 188, 252, 283
and 100% renewable household 293,
300
and Architecture 2030 Challenge 268
challenges to 271
costs of 92
NREL 276–278, 283
orientation/footprint factors 145, 147,
148, 245, 266, 277
and renovation standards 89–90, 256
ZEB 265–267, 274
businesses and renewable energy 12, 75,
256–257, 258
Butoni 101

C40 Initiative 88
California (US) 17, 18, 38, 193, 194,
213, 226, 227
Sonoma County case study 278–279
Canada 44, 88, 213, 231, 232
Canary Islands see El Hierro
cap-and-trade arrangements 12, 14
Caparroso (Spain) 127, 129
Cape Light (Massachusetts, US) 29
carbon capture and storage see CCS
carbon credits 20
carbon dioxide (CO$_2$) see greenhouse
gases
carbon emissions 1–2
carbon-neutrality 7
carbon offset schemes 12, 13
carbon sequestration 2, 12
carbon tax 14–15
carbon trading 11, 21, 26
see also ET